PRINCIPLES AND APPLICATIONS OF EMULSION POLYMERIZATION

PRINCIPLES AND APPLICATIONS OF EMULSION POLYMERIZATION

Chorng-Shyan Chern

WILEY
A JOHN WILEY & SONS, INC., PUBLICATION

Published by John Wiley & Sons, Inc., Hoboken, New Jersey
Published simultaneously in Canada

Library of Congress Cataloging-in-Publication Data:
Chern, Chorng-Shyan.
Principles and applications of emulsion polymerization / by Chorng-Shyan Chern.
p. cm.
Includes index.
ISBN 978-0-470-12431-4 (cloth)
1. Emulsion polymerization. I. Title.
QD281.P6C3927 2008
668.9′2—dc22

2008003740

10 9 8 7 6 5 4 3 2 1

CONTENTS

PREFACE

Emulsion polymerization is a unique chemical process that has been widely used to manufacture a variety of latex products for numerous applications. Some important examples include synthetic rubbers, adhesives, binders, caulks and sealants, trade paints, industrial coatings, printing inks and overprint varnishes, thermoplastics, emulsion aggregation toners, immunoassay products based on the affinity interaction between the ligand-containing latex particles and the target biomolecules, and monodisperse polymer particles for fine instrument calibration standards. The favorable atmosphere for environmentally friendly emulsion polymers is the major driving force for the rapid advancement of this green technology, though these water-based products have limitations in nature and perhaps they can never achieve the excellent performance properties offered by the solvent-based counterparts. In recent years, the continuously soaring crude oil price (reaching 100 U.S. dollars per barrel in the first quarter of 2008) makes these latex products compete more effectively with solvent-borne polymer systems in a variety of markets. The current trend clearly indicates that emulsion polymerization is an important field that deserves more research and development resources.

This technique, primarily based on (but not strictly limited to) conventional free radical polymerization mechanisms, continues to attract the attention of scientists and industrial professionals since the first introduction of styrene–butadiene copolymers and polyvinyl acetate for latex paints around 1946–1950. Considerable progress in the fundamental understanding of emulsion polymerization mechanisms and kinetics has been made since, not to mention many innovative emulsion polymers successfully developed in industrial laboratories. Extensive theoretical and experimental investigations and industrial product development efforts continue to advance our knowledge about the general features of emulsion polymerization mechanisms and kinetics. However, at present, some key points at issue such as particle nucleation and growth mechanisms and transport of free radicals simultaneous with chemical reactions occurring in the heterogeneous emulsion polymerization systems are still not completely understood. Furthermore, the issue of the colloidal stability of latex particles (i.e., the major reaction loci) may change the concentration of particles during polymerization, and this scenario makes the situation even more complicated. Besides the evolution of latex particles, determination of the average number of free radicals per particle is another task extremely difficult to undertake. These controversial, yet very important, subjects still remain a great challenge to those who are involved in this multidisciplinary research area.

In addition to the fundamental aspects of polymer chemistry and physics, a researcher dealing with emulsion polymerization must possess some basic knowledge of colloidal and interfacial phenomena, reaction kinetics, transport phenomena, thermodynamics, and polymer reaction engineering in order to effectively design and control latex products with desirable performance properties. Therefore, this book is aimed at providing comprehensive descriptions of conventional and surfactant-free emulsion polymerizations. These two reaction systems have been the most widely studied and employed in the plant production. The book also intends to provide a fundamental insight into some important features of the unique miniemulsion and microemulsion polymerization systems, which are expected to play an important role in emerging markets. Fundamental and quantitative interpretation of the polymerization mechanisms and kinetics involved in the heterogeneous reaction systems are the primary focuses of this volume. These subjects of vital importance, as reflected in a very large number of journal publications in the last half a century, enable the reader to quickly grasp the key reaction parameters that control the rate of polymerization and the particle size and molecular weight of the resultant emulsion polymers. Another goal is to provide introductory information on the colloidal phenomena related to emulsion polymerization and some industrial applications, common industrial emulsion polymerization processes (primarily semibatch and continuous reaction systems), latex particle morphology dealing with various types of multiphase polymer particles, and some important end-use properties of latex products.

This reference book or textbook is devoted to updating the current development of knowledge of emulsion polymerization. The author also endeavors to incorporate balanced fundamental and applied aspects of various emulsion polymerization processes into this work. This volume is particularly designed for research workers (such as chemists, chemical engineers, materials scientists, and physicists), technical service personnel, professors, and upper-level undergraduate and graduate students. It serves as an introduction to this important field and as a bridge to the more specialist-oriented books, which are available in the marketplace.

This book is dedicated to my Ph.D. thesis advisor, Professor Gary W. Poehlein, who introduced me to this challenging, fascinating research field in 1980s; to my parents, who inspired me to study in early years; and to my family (my beloved wife, Yue-Huan, and two lovely children, Andy and Angela). The author also expresses deep gratitude to Professor Poehlein for critically reviewing the manuscript and for the invaluable comments and thorough discussion to help improve the quality of this book. The assistance of the staff at John Wiley & Sons—especially the Consulting Editor, Dr. E. H. Immergut—throughout this work is gratefully acknowledged. This challenging project would not have been possible without all your encouragement and full support.

The author would like to conclude this Preface with a verse in the Bible (Psalm 19:1):

> The heavens declare the glory of God; and the firmament sheweth his handywork.

Without exception, an emulsion polymerization system, a world situated between the atomic/molecular level and the macro level, is subject to God's words, and it also responds to the ancient poet's praise with one accord. What a wonderful creation of God Almighty that is!

CHORNG-SHYAN CHERN

Taipei, Taiwan
April 2008

1

INTRODUCTION

Emulsion polymerization involves the reaction of free radicals with relatively hydrophobic monomer molecules within submicron polymer particles dispersed in a continuous aqueous phase. Nevertheless, this unique polymerization process that is heterogeneous in nature exhibits very different reaction mechanisms and kinetics compared to bulk or solution free radical polymerization. Surfactant is generally required to stabilize the colloidal system; otherwise, latex particles nucleated during the early stage of polymerization may experience significant coagulation in order to reduce the interfacial free energy. This feature may also come into play in determining the number of reaction loci (i.e., polymer particles) available for the consumption of monomer therein. The objective of this chapter is therefore to provide readers with an overview of those subjects such as the free radical polymerization mechanisms and kinetics, the general features of emulsion polymerization, the role of surfactants in emulsion polymerization, and the importance of colloidal stability that is sometimes ignored in this research area.

1.1 FREE RADICAL POLYMERIZATION

1.1.1 Free Radical Polymerization Mechanisms

Free radical polymerization of vinyl monomers containing carbon–carbon double bonds has been widely used in industry to manufacture a variety of polymeric materials such as low-density polyethylene, polystyrene, polyvinyl

Principles and Applications of Emulsion Polymerization, by Chorng-Shyan Chern

chloride, polyvinyl acetate, acrylic polymers, and synthetic rubbers, which can be accomplished in bulk, solution, suspension, or emulsion processes. The generally accepted free radical polymerization mechanism involves three kinetic steps in sequence, namely, initiation, propagation, and termination [1, 2].

Initiation

$$I \xrightarrow{k_d} 2R^* \tag{1.1}$$

$$R^* + M \xrightarrow{k_i} P_1^* \tag{1.2}$$

Propagation

$$P_1^* + M \xrightarrow{k_p} P_2^* \tag{1.3}$$

$$P_2^* + M \xrightarrow{k_p} P_3^* \tag{1.4}$$

$$P_n^* + M \xrightarrow{k_p} P_{n+1}^* \tag{1.5}$$

Termination

$$P_m^* + P_n^* \xrightarrow{k_{tc}} P_{m+n}^* \tag{1.6}$$

$$P_m^* + P_n^* \xrightarrow{k_{t,d}} P_m + P_n \tag{1.7}$$

where I, R*, M, P_n^* ($n = 1, 2, 3, \ldots$), and P_n represent the initiator, initiator radical, monomer, free radicals with n monomeric units, and dead polymer chains with n monomeric units, respectively. The kinetic parameters k_d, k_i, k_p, k_{tc}, and k_{td} are the thermal decomposition rate constant for the initiator, the initiation rate constant for the primary radical, the propagation rate constant for the reaction between one free radical with n monomeric units and one monomer molecule, the combination termination rate constant, and the disproportionation termination rate constant for the reaction between two free radicals, respectively.

The above three-reaction mechanism reflects its characteristic chain addition polymerization; the rate of consumption of monomer is relatively slow, but the molecular weight of polymer builds up rapidly, as shown schematically in Figure 1.1.

Chain transfer reactions are also a part of the free radical reaction system. These reactions, as the name implies, transfer the radical activity from a growing chain to another species such as monomer, polymer, initiator, solvent, or a deliberately added chain transfer agent. For example, chain transfer of a propagating radical to monomer or polymer can be represented as follows:

$$P_n^* + M \xrightarrow{k_{tr,m}} P_n + P_1^* \tag{1.8}$$

$$P_n^* + P_m \xrightarrow{k_{tr,p}} P_n + P_m^* \tag{1.9}$$

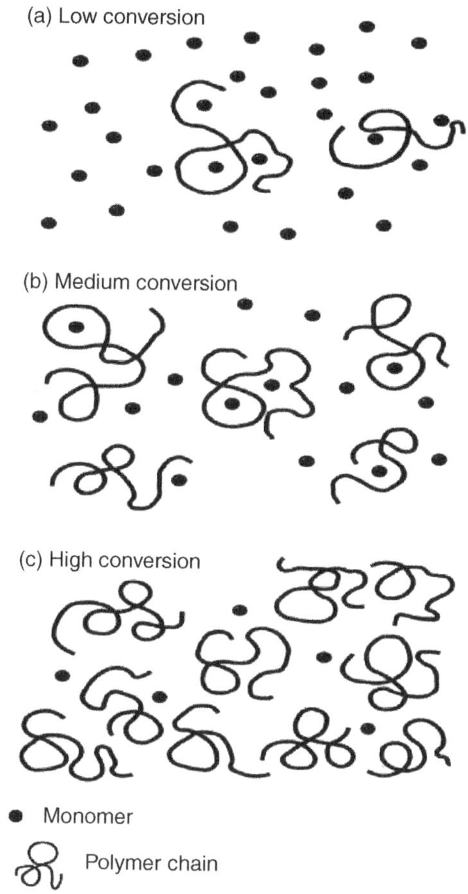

Figure 1.1. A schematic model for free radical polymerization at different levels of monomer conversion.

where $k_{tr,m}$ and $k_{tr,p}$ are the rate constants for the chain transfer reaction of a propagating radical with monomer and polymer, respectively. Both P_1^* and P_m^* may reinitiate the free radical chain polymerization to form linear and branched polymer chains, respectively, or participate in the termination reactions.

1.1.2 Free Radical Polymerization Kinetics

Assuming that the concentration of free radicals remains relatively constant during polymerization (the pseudo-steady-state assumption), the rate of polymerization (R_p) for bulk or solution polymerization can be expressed as

$$R_p = k_p[\mathrm{M}](fk_d[\mathrm{I}]/k_t)^{1/2} \tag{1.10}$$

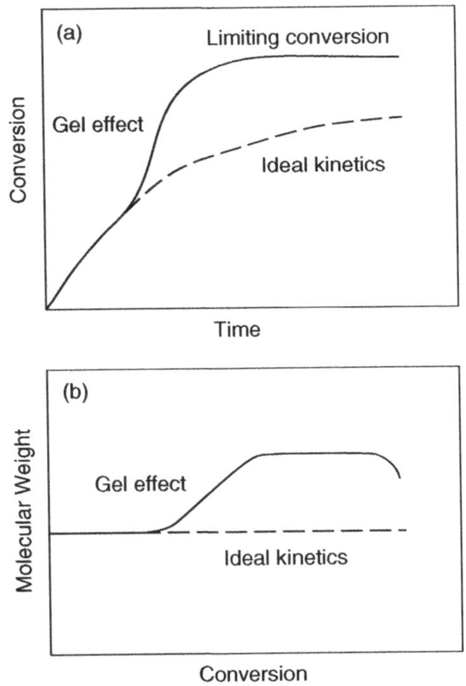

Figure 1.2. **(a)** Monomer conversion as a function of time. **(b)** Weight-average molecular weight as a function of conversion for free radical polymerization.

where f is the initiator efficiency factor and [M] and [I] are the concentrations of monomer and initiator, respectively. It should be noted that the term $(fk_d[\mathrm{I}]/k_t)^{1/2}$ represents the concentration of free radicals. At low monomer conversions, the rate of polymerization can be adequately predicted by Eq. (1.10). The conversion first increases and then gradually levels off with the progress of polymerization (Figure 1.2). However, for many free radical reactions, after a certain conversion, the termination rate constant (k_t) becomes chain length dependent due to the influence of diffusion of free radicals on the bimolecular termination reaction. Under these circumstances, k_t decreases significantly with increasing conversion, thereby leading to the severely retarded bimolecular termination reaction and then autoacceleration of the polymerization rate. This is termed the gel effect or Trommsdorff effect [3–6]. In general, k_p is relatively independent of the chain length of P_n^* because the rather mobile monomer predominates in the propagation reaction. Nevertheless, when the reaction temperature is below the T_g of the polymerizing medium, the propagation reaction may also become diffusion-controlled at very high conversions. Thus, k_p decreases continuously toward the end of polymerization and complete conversion cannot be achieved (termed the limiting conversion) [5, 6]. Mechanistic models based on the concept of free volume adequately describe the key features of the diffusion-controlled polymer reactions [3–6]. These peculiar kinetic phenomena are shown in Figure 1.2a.

The kinetic chain length (υ) can be calculated according to the following equation:

$$\upsilon = k_p[\mathrm{M}]/\left[2(fk_dk_t[\mathrm{I}])^{1/2}\right] \tag{1.11}$$

It should be noted that υ is inversely proportional to $[\mathrm{I}]^{1/2}$, and thus any attempt to increase the rate of polymerization (i.e., to shorten the batch cycle time) by increasing the concentration of initiator inevitably results in polymer with a shorter chain length. Equation (1.11) predicts that molecular weight will change with the ratio $[\mathrm{M}]/[\mathrm{I}]^{1/2}$ during the course of the reaction if the rate constants are indeed constant. By contrast, molecular weight is strongly dependent on conversion due to the diffusion-controlled polymer reactions (Figure 1.2b).

1.2 EMULSION POLYMERIZATION

1.2.1 Conventional Emulsion Polymerization

Emulsion polymerization, which is a heterogeneous free radical polymerization process, involves emulsification of the relatively hydrophobic monomer in water by an oil-in-water emulsifier, followed by the initiation reaction with either a water-soluble initiator (e.g., sodium persulfate (NaPS)) or an oil-soluble initiator (e.g., 2,2′-azobisisobutyronitrile (AIBN)) [7–15]. This polymerization process was first commercialized in the early 1930s, and since then it has been widely used to produce environmentally friendly latex products with a variety of colloidal and physicochemical properties. If desired, these water-based polymer dispersions can be readily converted into bulk resins. Some representative monomers used to synthesize emulsion polymers include ethylene, butadiene, styrene, acrylonitrile, acrylate ester and methacrylate ester monomers, vinyl acetate, and vinyl chloride. Because the compatibility between the polymer produced and water is very poor, an exceedingly large oil–water interfacial area is generated as the particle nuclei form and grow in size with the progress of the polymerization. Thus, effective stabilizers such as ionic surfactants, nonionic surfactants, or protective colloids (e.g., hydroxyethyl cellulose, polyvinyl alcohol, polyvinyl pyrrolidone and dextrin), which can be physically adsorbed or chemically incorporated onto the particle surface, are generally required to prevent the interactive latex particles from coagulation. Satisfactory colloidal stability can be achieved via the electrostatic stabilization mechanism [16], the steric stabilization mechanism [17, 18], or both. Latex products comprise a very large population of polymer particles (~10^1–10^3 nm in diameter) dispersed in the continuous aqueous phase. Although the performance properties of most of the water-based polymers are often inferior to their solvent-based counterparts, there has been an even stronger demand for these environmentally friendly products in recent years. The continuously increased pressure for environmental protection and a hike in crude oil price are the primary factors responsible for this constant trend. The major competitive technologies in the marketplace include high-solids solvent-based polymers, UV coatings, and powder coatings.

To gain an insight into emulsion polymerization mechanisms and kinetics is a must in the development of quality products that fulfill customers' requirements. Unfortunately, this heterogeneous polymerization system is a very complex process because nucleation, growth, and stabilization of polymer particles are controlled by the above-mentioned free radical polymerization mechanisms along with several colloidal phenomena. There is no doubt that the most striking feature of emulsion polymerization is the segregation of free radicals among the discrete monomer-swollen polymer particles during the reaction. This unique behavior greatly reduces the probability of bimolecular termination of free radicals, thereby resulting in a faster polymerization rate and, in the absence of significant chain transfer reactions, producing a polymer with a higher molecular weight. This advantageous characteristic of emulsion polymerization cannot be achieved simultaneously in bulk or solution polymerization (see Section 1.1). Although the particle nucleation period is relatively short, formation of particle embryos during the early stage of polymerization plays an important role in determining the final latex particle size and particle size distribution, and it also affects quality control and application properties (e.g., colloidal stability, rheology, and film formation) of latex products. How to effectively control the particle nucleation process then represents a very challenging task to those who are involved in this fascinating research area. Beyond the particle nucleation stage, transport of monomer molecules, free radicals, and stabilizer molecules to the growing particles and partition of these reagents among the continuous aqueous phase, emulsified monomer droplets (monomer reservoir), monomer-swollen polymer particles (primary reaction loci), and oil–water interface are the key factors that govern the particle growth stage. The colloidal properties of latex products are of great importance from both academic and industrial points of view. Some representative properties include the particle size and particle size distribution, particle surface charge density (or zeta potential), particle surface area covered by one stabilizer molecule, conformation of the hydrophilic polymer physically adsorbed or chemically coupled onto the particle surface, type and concentration of functional groups on the particle surface, particle morphology, optical and rheological properties, and colloidal stability. Table 1.1 summarizes the primary characteristics of bulk free radical polymerization and emulsion polymerization.

1.2.2 Emulsion Polymerization Processes

Three types of reactors that are used to produce latex products and will be briefly discussed here are the batch reactor, semibatch reactor, and continuous

Table 1.1. Primary Characteristics of Bulk Free Radical Polymerization and Emulsion Polymerization

Parameter	Bulk Polymerization	Emulsion Polymerization
Reaction system	Homogeneous	Heterogeneous
R_p/MW[a]	Slow/high or fast/low	Fast/high
Temperature control	Poor	Good
Viscosity of medium	High	Low

[a] R_p and MW are the rate of polymerization and molecular weight of polymer, respectively.

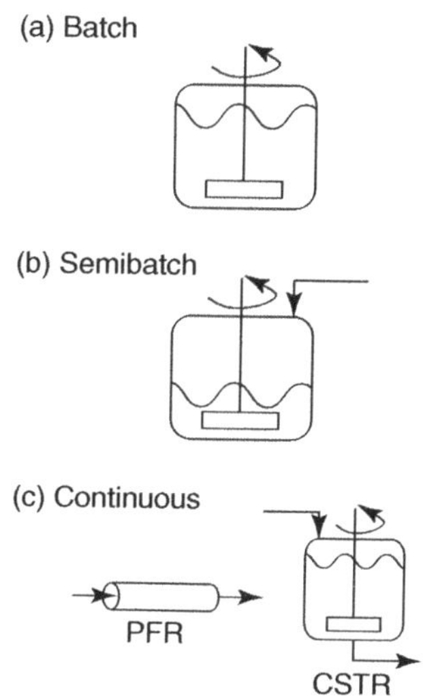

Figure 1.3. Three types of reactors commonly used to produce latex products: **(a)** batch reactor, **(b)** semibatch reactor, and **(c)** continuous reactor.

reactor (Figure 1.3). The reader is referred to an excellent book dealing with these three polymerization systems [19]. Batch emulsion polymerization is generally used in the laboratory to study reaction mechanisms, develop new latex products, and obtain kinetic data for process development and reactor scale-up. To the best of the author's knowledge, only very few latex products are manufactured in large-scale batch polymerization systems. As expected, severe problem in controlling the reaction temperature is often experienced because free radical polymerization is highly exothermic in nature and the heat transfer capacity is rather limited due to the very small surface-to-volume ratio of large-scale batch reactors. Most of the commercial latex products are thus produced by semibatch or continuous reaction systems. Continuous reaction systems include plug flow reactors (PFR), continuous stirred tank reactor trains (CSTR) and any combinations of PFR and CSTR. One major difference among the above polymerization processes is the residence time distribution of the growing particles within the reactor system. The broadness of residence time distribution in decreasing order is a single CSTR > semibatch > batch. As a result, the broadness of the resultant particle size distribution in decreasing order is single CSTR > semibatch > batch. It should be noted that PFR behaves much like the batch reactor in terms of residence time distribution, provided that the length of PFR is long enough. Nevertheless, the very different fluid dynamics patterns observed in the PFR and batch reactor may have an influence on the properties of latex products to some extent. The rate of polymerization generally follows the following trend: batch > semibatch > CSTR. In addition, the versatile semibatch and continuous emulsion polymerization systems offer the operational flexibility to prepare latex products with controlled particle size distribution, polymer composition, and particle morphology. This may serve as an effective tool for chemists to design specialty emulsion polymers exhibiting performance properties that meet the sometimes contradictory requirements of customers. For the continuous emulsion polymerization systems that are only suitable for large-volume latex products, more efforts should be devoted to the development and understanding of different reactor designs and operating procedures, especially related to characteristics of latex products and start-up and product changeover strategies. Only a small number of professional journal papers dealing with these two important industrial processes are available in the open literature. More research efforts are required to further advance the semibatch and continuous emulsion polymerization technology. For those who are interested in the previous studies on semibatch and continuous emulsion polymerizations, refer to those cited in the review articles [20–25].

1.2.3 Miniemulsion Polymerization

Miniemulsion, microemulsion, and classical emulsion polymerizations show quite different particle nucleation and growth mechanisms and kinetics. This is primarily attributed to the different initial conditions (i.e., the conditions immediately before the start of polymerization) that have a profound influence on the subsequent particle nucleation and growth mechanisms. In the conventional emulsion polymerization, the most widely accepted particle nucleation mechanisms include micellar nucleation [26–31] and homogeneous nucleation [32–36]. Emulsified monomer droplets (>10 μm in diameter) generally do not contribute to particle nucleation to any appreciable extent due to their very small droplet surface area (Figure 1.4a). However, after intensive homogenization, submicron monomer droplets containing a hydrophobic, low-molecular-weight compound [e.g., hexadecane (HD) or cetyl alcohol (CA)] may become the predominant particle nucleation loci if the total monomer droplet surface area becomes large enough to compete effectively with the continuous aqueous phase, in which particle nuclei are generated to capture radicals (i.e., monomer droplet nucleation). The hydrophobic, low-molecular-weight species is used as a costabilizer to impart the osmotic pressure effect

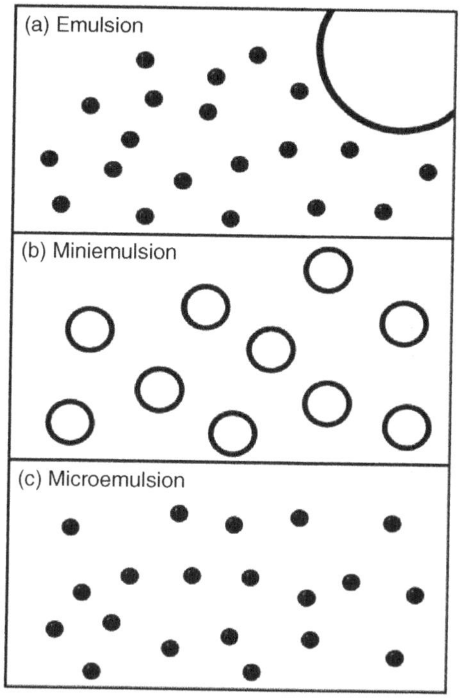

Figure 1.4. Initial conditions for **(a)** the conventional emulsion polymerization, **(b)** miniemulsion polymerization with the surfactant concentration lower than its critical micelle concentration, and **(c)** microemulsion polymerization. ○ (>10^4 nm in diameter for the conventional emulsion polymerization and <10^3 nm in diameter for miniemulsion polymerization) and ● (~10^0 nm in diameter) represent emulsified monomer droplets and monomer-swollen micelles, respectively.

to the emulsion system to retard the diffusion of monomer molecules from smaller droplets to larger ones (i.e., the degradation of monomer droplets or Ostwald ripening effect). As a result, kinetically stable miniemulsion droplets are produced for the subsequent free radical polymerization (Figure 1.4b). This unique technique has been termed the miniemulsion polymerization [37–42], and new industrial applications based on miniemulsion polymerization have been continuously developed. One major potential advantage of miniemulsion polymerization is the ability to utilize highly water-insoluble monomers and other reagents because with droplet polymerization those ingredients do not need to be transported through the aqueous phase.

1.2.4 Microemulsion Polymerization

Unlike the conventional milky white emulsion, the transparent or translucent reaction system comprising microemulsion droplets is thermodynamically stable in nature, and these tiny droplets (~1–10 nm in diameter) have an extremely large oil–water interfacial area (~$10^5\, m^2\, dm^{-3}$), as shown schematically in Figure 1.4c. An anionic surfactant (e.g., sodium dodecyl sulfate (SDS)) in combination with a cosurfactant (e.g., 1-pentanol) is the most popular stabilization package. Incorporation of amphipathic 1-pentanol into the adsorbed layer of SDS around an oil droplet greatly reduces the electrostatic repulsion force between two anionic SDS molecules, minimizes the oil–water interfacial tension, and enhances the flexibility of interfacial membrane. All these synergistic factors promote the spontaneous formation of a transparent one-phase microemulsion. Relatively stable polymer particles (~10 nm in diameter) consisting of only a few polymer chains per particle are produced, and therefore the resultant polymer molecular weight is very high (~10^6–$10^7\, g\, mol^{-1}$) [43–47]. This cannot be achieved readily by the conventional emulsion polymerization or miniemulsion polymerization. In addition, the particle nucleation and growth mechanisms and kinetics associated with microemulsion polymerization are quite different from those of emulsion and miniemulsion polymerization systems. Research interests in such a polymerization technique have grown rapidly since the 1980s because of its potential applications in the preparation of fine latex particles, ultrahigh-molecular-weight water-soluble polymers (flocculants), novel porous materials, polymeric supports for binding metal ions, conducting polymers, colloidal particles containing various functional groups for the biomedical field, and transparent colloidal systems for photochemical and other chemical reactions [45].

In summary, there has been a tremendous effort devoted to the fundamental aspects of emulsion polymerization mechanisms, kinetics and processes since the early twentieth century. Representative review or journal articles concerning the conventional emulsion polymerization can be found in literature [20–25, 48–60]. The research areas related to both miniemulsion [42, 61–64] and microemulsion [44–47, 65] polymerizations have received increasing interest recently.

1.2.5 Inverse Emulsion Polymerization

Inversion emulsion polymerization involves the dispersion and then polymerization of hydrophilic monomers, normally in aqueous solution, in a nonaqueous continuous phase. The emulsifier systems primarily based on the steric stabilization mechanism (see Section 1.3.3) are quite different from those of the more conventional oil-in-water emulsion polymerization processes. This is simply because the electrostatic stabilization mechanism (see Section 1.3.2) is not effective in stabilizing inverse emulsion polymerization comprising an aqueous disperse phase and a nonaqueous continuous phase with a very low dielectric constant. The unique anionic surfactant bis(2-ethylhexyl) sulfosuccinate (trade name: Aerosol OT) that can be dissolved in both oil and water has been widely used in the stabilization of inverse emulsion or microemulsion polymerization systems. Both oil-soluble and water-soluble initiators have been used to initiate the free radical chain polymerization in this water-in-oil type of colloidal systems.

The inverse emulsion polymerization mechanisms and kinetics can be found in the literature [10, 66–68]. The area of inverse emulsion polymerization has not been studied extensively, except perhaps for the inverse microemulsion polymerization of acrylamide. The most important applications for these acrylamide-based products are as polymeric flocculants in water treatment. The two major advantages of this polymerization process are the very high polymer molecular weight and a colloidal system that results in rapid dissolution of the polymer in water.

1.3 COLLOIDAL STABILITY

1.3.1 A Critical but Often Ignored Issue

The product obtained from (conventional) emulsion polymerization is a colloidal dispersion comprising a very large population of submicron hydrophobic polymer particles dispersed in the continuous aqueous phase. This colloidal system is not thermodynamically stable because of the incompatibility between polymer and water (i.e., the very low solubility of polymer in water) in nature. As a matter of fact, the fate of most common latex products is the coagulation of polymer particles in order to minimize the particle–water interfacial area. Moreover, the monomer-swollen particles may even lose their colloidal stability and flocculate with one another in the course of emulsion polymerization. This will inevitably make the particle nucleation and growth mechanisms more complicated.

The level of surfactant used in emulsion polymerization is generally kept at a minimum for the manufacture of polymeric materials with excellent water resistance and adhesion properties. However, polymer particles that are inadequately stabilized by stabilizers (anionic and nonionic surfactants and protective colloids) may lose their colloidal stability, and coagulation of these unstable particles occurs upon aging. This represents an extremely challenging task for those who deal with the surfactant-free emulsion polymerization technique. Therefore, the shelf life of latex products is an important issue for successful product development. Furthermore, intensive coagulation of the latex particles to form filterable solids and scraps adhering to the reactor wall and agitator could become a serious problem in plant production. This annoying production problem not only increases the cost significantly but also makes the task of quality control much more difficult. Thus, the colloidal stability issue that has sometimes been ignored must be addressed from both the theoretical and practical points of view.

1.3.2 Electrostatic Interactions

Adequate colloidal stability of the emulsion polymerization system can be achieved by the adsorption of surfactants and protective colloids onto the growing polymer particle surfaces with the progress of polymerization. For example, the interactions between two negatively charged colloidal particles originating from anionic surfactant molecules adsorbed on their particle surfaces can be described schematically in Figure 1.5 [16, 69]. The total potential energy of interaction is the resultant of the competitive van der Waals attraction force and electrostatic repulsion force between the two adjacent particles. The driving force for coagulation of colloidal particles to take place is van der Waals attraction force. The positive and negative signs of the total potential energy of interaction represent the net repulsive interaction and the net attractive interaction between two particles, respectively.

The total potential energy of interaction approaches zero, and this implies that the interactions are insignificant when the distance of separation between two particles is very large, as shown in Figure 1.5. As the distance of separation decreases, the attractive van der Waals force tends to pull the two particles together. Further decreasing the distance of separation, the pair of particles

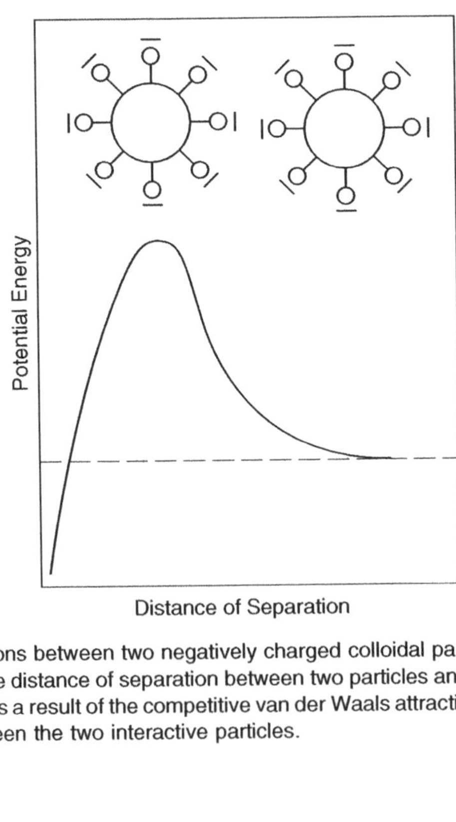

Figure 1.5. Interactions between two negatively charged colloidal particles. The abscissa and ordinate represent the distance of separation between two particles and potential energy barrier against coagulation as a result of the competitive van der Waals attraction force and electrostatic repulsion force between the two interactive particles.

starts to experience the electrostatic repulsion force to some extent. At close approach, the total potential energy barrier (if high enough) may oppose the continued approach of the interactive particles. The primary maximum observed in the total potential energy of interaction versus distance of separation curve represents an activation energy that must be overcome for coagulation to take place. It is dependent on the constituent materials in the disperse phase, the surface charge density (or zeta potential) of colloidal particles, the valency of counterions, and the concentration of electrolytes.

The DLVO theory predicts that decreasing the surface charge density, increasing the valency of counterions, and increasing the concentration of electrolytes tend to depress the primary maximum and thus reduce the colloidal stability [16, 69]. Another useful rule of thumb, the Shultz–Hardy rule, suggests that the critical coagulation concentration, defined as the concentration of counterions at which point visible aggregates start to form in the colloidal dispersion, is inversely proportional to the valency of counterions to the sixth power [70, 71]. This strongly suggests that the addition of counterions with a higher valency should be avoided during emulsion polymerization or in subsequent formulation work. On the other hand, a simple method based on the Shultz–Hardy rule can be developed to evaluate the effectiveness of the electrostatic stabilization of a particular colloidal dispersion.

1.3.3 Steric Interactions

In addition to the electrostatic stabilization mechanism, latex particles can be stabilized by adsorption of hydrophilic polymer chains on their particle surfaces. The physically adsorbed or chemically grafted polymer chains surrounding the colloidal particles and extending themselves into the continuous aqueous phase serve as a steric barrier against the close approach of the pair of particles. In this manner, coagulation of latex particles can be prevented via the steric stabilization mechanism [17, 18]. Typical nonionic surfactants and surface-active, nonionic block copolymers are quite effective in imparting such a steric stabilization effect to colloidal dispersions.

A schematic representation of the steric stabilization mechanism is shown in Figure 1.6. Considering the space of interaction between two hairy colloidal particles as the control volume, these particles are well separated and there is no overlap of adsorbed polymer layers initially (Figure 1.6a). In this case, the adsorbed polymer chain has the maximum number of conformations near the particle surface, and therefore the corresponding entropy is relatively large. Under these circumstances, no appreciable interactions between the pair of particles can be detected. However, at close approach, the concentration of polymer increases, and therefore the number of polymer chain confirmations (i.e., entropy) decreases significantly due to the overlap of adsorbed polymer layers (Figure 1.6b). As a result, the change of entropy (i.e., the entropy of the final state (Figure 1.6b) minus that of the initial state

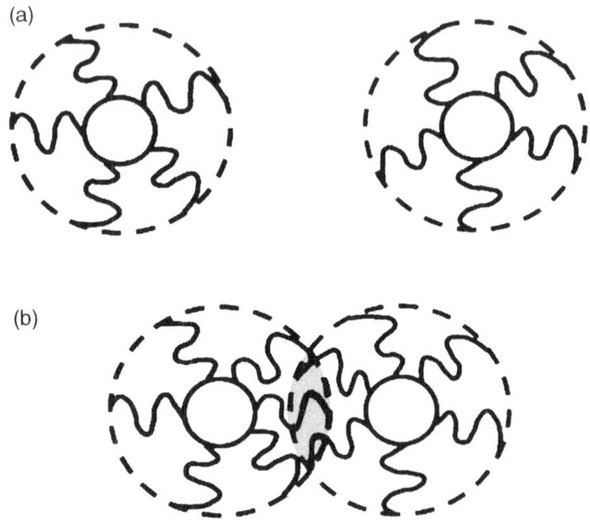

Figure 1.6. Interactions between two hairy colloidal particles: **(a)** Two particles with a relatively large distance of separation in the absence of overlap of adsorbed polymer layers and **(b)** two particles at close approach in the presence of overlap of adsorbed polymer layers.

(Figure 1.6a)) is negative. This implies that this process is not thermodynamically feasible, and repulsive steric interactions are established accordingly to pull the pair of particles apart. In general, the degree of steric interactions between two hairy particles is governed by the surface polymer concentration, the polymer chain length (or the adsorbed polymer layer thickness), and temperature. The primary features of models describing the steric stabilization mechanism include (a) a strong effect of temperature on the repulsive steric interactions, (b) a rapidly increased steric interactions with increasing polymer concentration in the adsorbed polymer layer, and (c) an increased steric interactions with decreasing distance of separation [17, 18, 70, 71]. The addition of a solvent or other ingredient that could cause polymer desorption from the particle surfaces could clearly destabilize a sterically stabilized system.

1.3.4 Mechanical Stability

The shear force used to provide the heterogeneous reaction system with efficient mixing and heat transfer also has an influence on the colloidal stability of latex particles during polymerization. The shear force is linearly proportional to the velocity gradient, the viscosity of the continuous phase, and the square of the colloidal particle size. This relationship can be used to determine where the stability or instability boundaries are located for a particular colloidal dispersion (e.g., the mechanical stability map for a particular colloid dispersion as a function of shear rate and zeta potential) [70, 71]. It is also interesting to note that at constant velocity gradient and viscosity, the mechanical stability of larger colloidal particles is more sensitive to the hydrodynamic force than that of smaller particles.

The above discussion reflects an important fact that a variety of interfacial phenomena play important roles in heterogeneous reaction systems such as emulsion polymerization, but somehow this scientific discipline has not been studied as extensively of the reaction systems over the years. Therefore, the reader is encouraged to become familiar with the most basic concepts of colloid and interface science.

1.4 SOME PERFORMANCE PROPERTIES FOR INDUSTRIAL APPLICATIONS

One distinct advantage of the emulsion polymerization technique is that latex products are often used directly without prior separation of polymer from water. For example, coatings formulated primarily with emulsion polymers are essential to the beauty and protection of many objects such as houses, furniture, leathern products, and packaging materials. The performance properties of emulsion polymers of major interest to this section include rheology and film formation related to the colloidal phenomena. These performance properties play a crucial role in determining the ultimate mechanical properties of the polymeric films.

1.4.1 Rheology

Rheology deals with the deformation or flow of materials subjected to forces [71, 72], and it has an influence on (a) mixing and heat transfer during emulsion polymerization, (b) transfer of latex products, (c) storage, handling, and mixing of emulsion polymers, cosolvents, and additives to manufacture water-based coatings, (d) application of the finished coatings products to a variety of substrates (e.g., brush coating, roll coating, and spray coating), and (e) film formation. For industrial coatings, satisfactory rheological properties are necessary to ensure the formation of polymeric films with excellent appearance and physical and chemical properties. The rheological characteristics of materials can be described by the dimensionless Deborah number De, defined as the ratio of the characteristic time τ to the experimental time t_{exp}. Table 1.2 summarizes the conditions corresponding to three types of rheological behavior. It should be noted that colloidal dispersions have De values close to unity, and therefore they can exhibit viscoelastic behavior.

Table 1.2. Ranges of Deborah Number Corresponding to Three Types of Rheological Behavior

De	Rheological Behavior	Examples
<<1	Liquid-like	Water
~1	Viscoelastic	Latex paints
>>1	Solid-like	Steel

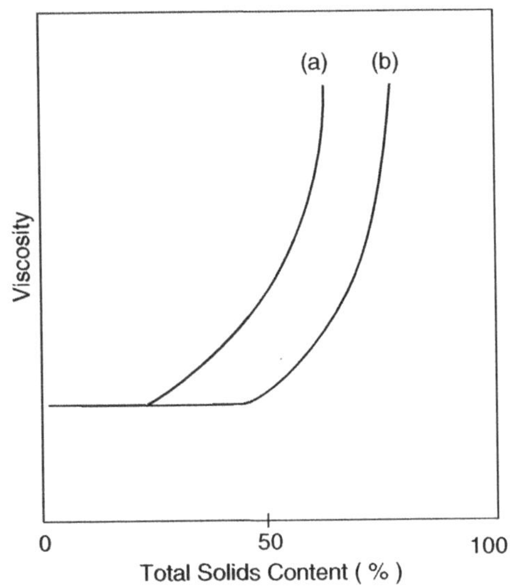

Figure 1.7. Viscosity of the colloidal dispersion versus total solids content profiles with different particle sizes: **(a)** small particle size and **(b)** large particle size.

The viscosity of a colloidal dispersion is a rheological property that measures the resistance to flow in response to the applied shear force. It is dependent on the hydrodynamic interactions between the particles and the continuous aqueous phase and interparticle interactions. The viscosity increases exponentially with increasing total solids content of the emulsion polymer, as shown schematically in Figure 1.7. This general feature can be described by the Mooney equation [73]:

$$\eta = \eta_0 \exp\{2.5\varphi/[1-(\varphi/\varphi^*)]\} \qquad (1.12)$$

where η and η_0 are the viscosity of the colloidal dispersion and the viscosity of the continuous aqueous phase, respectively. The parameters φ and φ^* represent the volume fraction of the colloidal particles and the critical volume fraction corresponding to infinite viscosity, respectively. For very dilute col-

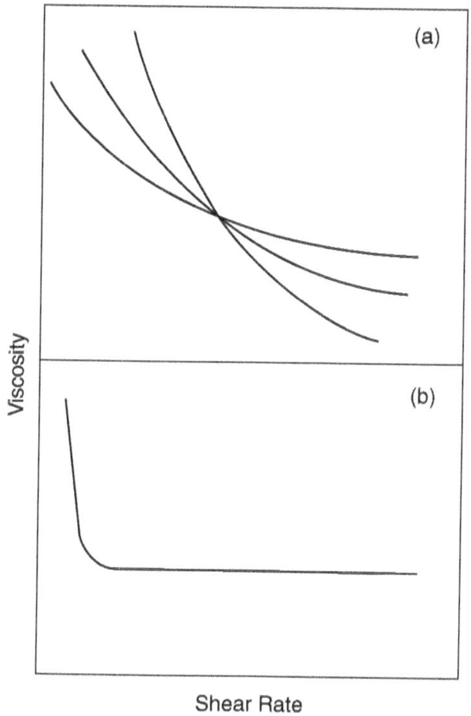

Figure 1.8. (a) Viscosity of the colloidal dispersion versus shear rate profiles that can be adjusted by a variety of rheology modifiers. **(b)** Viscosity versus shear rate curve for the rheology-controlled emulsion polymers.

loidal dispersions, Eq. (1.12) can be reduced to the well-known Einstein equation:

$$\eta = \eta_0(1 + 2.5\varphi) \tag{1.13}$$

The crowding effect is more pronounced for the dispersion with a smaller particle size (Figure 1.7a). At constant total solids content, the viscosity of the dispersion increases significantly with decreasing particle size. This is the major reason why it is very difficult to prepare latex products with high solids contents and small particle sizes simultaneously. High-solids-contents latexes are often comprised of broad particle size distributions, including bimodal systems. In extreme cases, the numerous small particles act as part of the continuous phase for flow of the large particles. In addition, the viscosity of a colloidal dispersion decreases with increasing temperature.

Normally, emulsion polymers show a shear-thinning behavior; viscosity decreases with increasing shear rate (Figure 1.8a). Furthermore, the shape of the viscosity versus shear rate curve can be adjusted by a variety of effective rheology modifiers to meet various application requirements of end-users. Typical rheology modifiers include cellulosic thickeners, alkali-soluble or -swellable thickeners, and advanced associative thickeners. It is also interesting to note that a Newtonian-like flow behavior illustrated in Figure 1.8b is observed for some latex products (e.g., a series of rheology-controlled emulsion polymers originally developed at S. C. Johnson Wax). The viscosity of this type of latex products first decreases rapidly and then levels off as the shear rate is increased.

1.4.2 Film Formation

After application, the liquid coating must be converted into a solid polymeric film (viscosity > 10^6 cps) in order to build up satisfactory performance properties (termed the film formation process) [74]. As water evaporates from a film of emulsion polymer, the distance of separation between the submicron particles continues to decrease and, ultimately, capillary tubes form. In a capillary tube, surface tension results in a force that tends to collapse the tube. Moreover, the smaller the diameter of the tube, the greater the destruction force. When the particles are so close to one another, the destruction force is strong enough to overcome the repulsion forces originating from either the electrostatic or steric interaction mechanism striving to push the neighboring particles apart. Coalescence of the particles to form a continuous film is thus possible.

To produce an integral film during the latter stage of the film formation process is primarily governed by the free volume of the coating materials. This is simply because polymer chains must be free to diffuse from particle to particle so that the individual particle boundaries disappear in the completely coalesced film. Above the glass transition temperature (T_g) of polymer, free volume increases (or viscosity decreases) significantly with increasing temperature. This then results in decreased viscosity and increased polymer chain mobility with temperature. In other words, the rate of coalescence is controlled by $T - T_g$, and it is much easier for film formation to occur when the temperature is much higher than the T_g. The following universal equation can be used to estimate the T_g required for a polymeric film to be "dry to touch" at a particular temperature T.

$$\ln \eta = 27.6 - \{40.2(T - T_g)/[51.6 + (T - T_g)]\} \tag{1.14}$$

For example, the T_g required for a polymeric film to be "dry to touch" (i.e., a solid-like film with $\eta = 10^6$ cps $= 10^3$ Pa s) at 25 °C is estimated to be −29 °C [74]. However, a polymeric film with $T_g = -29$ °C is kind of sticky, and it is only useful in the area of adhesives. According to Eq. (1.14), satisfactory film formation apparently cannot be achieved for common latex paints with T_g in the range 0–30 °C when applied at some temperature below 25 °C. To resolve this problem, suitable coalescing solvents can be incorporated into the latex paint formulations to lower the minimum film formation temperature (MFFT). After the film formation process is completed, coalescing solvent molecules can migrate toward the coating surface layer and then evaporate.

It should be pointed out that satisfactory film formation is a prerequisite for the coating materials to fully exhibit their excellent mechanical properties. Finally, the mechanical properties (e.g., hardness, flexibility, impact resistance, solvent resistance, abrasion resistance, post-formability, and adhesion) of the polymeric film are closely related to the T_g, structure, molecular weight, and viscoelasticity of polymers.

REFERENCES

1. G. Odian, *Principles of Polymerization*, 2nd ed., Wiley Interscience, New York, 1981, Chapter 3.
2. M. P. Stevens, *Polymer Chemistry: An Introduction*, 2nd ed., Oxford University Press, New York, 1990, Chapter 6.
3. S. K. Soh and D. C. Sundberg, *J. Polym. Sci.: Polym. Chem. Ed.* **20**, 1299 (1982).
4. S. K. Soh and D. C. Sundberg, *J. Polym. Sci.: Polym. Chem. Ed.* **20**, 1315 (1982).
5. S. K. Soh and D. C. Sundberg, *J. Polym. Sci.: Polym. Chem. Ed.* **20**, 1331 (1982).
6. S. K. Soh and D. C. Sundberg, *J. Polym. Sci.: Polym. Chem. Ed.* **20**, 1345 (1982).
7. F. A. Bovey, I. M. Kolthoff, A. I. Medalia, and E. J. Meehan, *Emulsion Polymerization*, Interscience Publishers, New York, 1965.
8. D. C. Blakely, *Emulsion Polymerization—Theory and Practice*, Applied Science, London, 1975.
9. V. I. Eliseeva, S. S. Ivanchev, S. I. Kuchanov, and A. V. Lebedev, *Emulsion Polymerization and Its Applications in Industry*, Consultants Bureau, New York and London, 1981.
10. J. Barton and I. Capek, *Radical Polymerization in Disperse Systems*, Ellis Horwood, New York, 1994.
11. R. G. Gilbert, *Emulsion Polymerization: A Mechanistic Approach*, Academic Press, London, 1995.
12. R. M. Fitch, *Polymer Colloids: A Comprehensive Introduction*, Academic Press, London, 1997.
13. P. A. Lovell and M. S. El-Aasser (Eds.), *Emulsion Polymerization and Emulsion Polymers*, John Wiley & Sons, West Sussex, 1997.
14. H. Y. Erbil, *Vinyl Acetate Emulsion Polymerization and Copolymerization with Acrylic Monomers*, CRC Press LLC, Boca Raton, FL, 2000.

15. A. van Herk, *Chemistry and Technology of Emulsion Polymerization*, Blackwell Publishing, Oxford, 2005.
16. E. J. W. Verwey and J. Th. G. Overbeek, *Theory of the Stability of Lyophobic Colloids*, Elsevier, Amsterdam, 1948.
17. T. Sato and R. Ruch, *Stabilization of Colloidal Dispersions by Polymer Adsorption*, Marcel Dekker, New York, 1980.
18. D. H. Napper, *Polymeric Stabilization of Colloidal Dispersions*, Academic Press, London, 1983.
19. F. J. Schork, P. B. Deshpande, and K. W. Leffew, *Control of Polymerization Reactors*, Marcel Dekker, New York, 1993, Chapter 3.
20. G. W. Poehlein and D. J. Dougherty, *Rubber Chem. Technol.* **50**, 601 (1977).
21. B. Li and B. W. Brooks, *Polym. Int.* **29**, 41 (1992).
22. J. Snuparek, *Prog. Org. Coat.* **29**, 225 (1996).
23. J. Gao and A. Penlidis, *Prog. Polym. Sci.* **27**, 403 (2002).
24. C. S. Chern, in *Encyclopedia of Surface and Colloid Science*, A. Hubbard (Ed.), Marcel Dekker, New York, 2002, pp. 4220–4241.
25. M. Nomura, H. Tobita, and K. Suzuki, *Adv. Polym. Sci.* **175**, 1 (2005).
26. W. D. Harkins, *J. Chem. Phys.* **13**, 381 (1945).
27. W. D. Harkins, *J. Chem. Phys.* **14**, 47 (1946).
28. W. D. Harkins, *J. Am. Chem. Soc.* **69**, 1428 (1947).
29. W. V. Smith, *J. Am. Chem. Soc.* **70**, 3695 (1948).
30. W. V. Smith and R. H. Ewart, *J. Chem. Phys.* **16**, 592 (1948).
31. W. V. Smith, *J. Am. Chem. Soc.* **71**, 4077 (1949).
32. W. J. Priest, *J. Phys. Chem.* **56**, 1977 (1952).
33. C. P. Roe, *Ind. Eng. Chem.* **60**, 20 (1968).
34. R. M. Fitch and C. H. Tsai, Particle formation in polymer colloids, III: Prediction of the number of particles by a homogeneous nucleation theory, in *Polymer Colloids*, R. M. Fitch (Ed.), Plenum Press, New York, 1971, p. 73.
35. R. M. Fitch and C. H. Tsai, Homogeneous nucleation of polymer colloids. IV: The role of soluble oligomeric radicals, in *Polymer Colloids*, R. M. Fitch (Ed.), Plenum Press, New York, 1971, p. 103.
36. R. M. Fitch, *Br. Polym. J.* **5**, 467 (1973).
37. J. Ugelstad, M. S. El-Aasser, and J. W. Vanderhoff, *J. Polym. Sci.: Polym. Lett. Ed.* **11**, 503 (1973).
38. J. Ugelstad, F. K. Hansen, and S. Lange, *Die Makromol. Chem.* **175**, 507 (1974).
39. D. P. Durbin, M. S. El-Aasser, G. W. Poehlein, and J. W. Vanderhoff, *J. Appl. Polym. Sci.* **24**, 703 (1979).
40. B. J. Chamberlain, D. H. Napper, and R. G. Gilbert, *J. Chem. Soc. Faraday Trans. 1* **78**, 591 (1982).
41. Y. T. Choi, M. S. El-Aasser, E. D. Sudol, and J. W. Vanderhoff, *J. Polym. Sci.: Polym. Chem. Ed.* **23**, 2973 (1985).
42. I. Capek and C. S. Chern, *Adv. Polym. Sci.* **155**, 101 (2001).
43. F. Candau, in *Encyclopedia of Polymer Science and Engineering*, H. F. Mark, N. M. Bikales, C. G. Overberger, and G. Mengers (Eds.), Wiley, New York, 1987, p. 718.

44. F. Candau, in *Polymerization in Organized Media*, C. M. Paleos (Ed.), Gordon and Breach, New York, 1992, Chapter 4, p. 215.
45. F. Candau, in *Handbook of Microemulsion Science and Technology*, P. Kumar and K. L. Mittal (Eds.), Marcel Dekker, New York, 1999, p. 679.
46. I. Capek, *Adv. Colloid Interface Sci.* **80**, 85 (1999).
47. I. Capek, *Adv. Colloid Interface Sci.* **82**, 253 (1999).
48. A. E. Alexander and D. H. Napper, Emulsion polymerization, *in Progress in Polymer Science*, D. P. Jensen (Ed.), Pergamon Press, Oxford, 1971, p. 145.
49. K. E. Barrett, *Dispersion Polymerization in Organic Media*, Wiley, New York, 1975.
50. J. Ugelstad and F. K. Hansen, *Rubber Chem. Technol.* **49**, 536 (1976).
51. J. W. Vanderhoff, *J. Polym. Sci. Polym. Symp.* **72**, 161 (1985).
52. Q. Wang, S. Fu, and T. Yu, *Prog. Polym. Sci.* **19**, 703 (1994).
53. T. R. Aslamazova, *Prog. Org. Coat.* **25**, 109 (1995).
54. K. Nagai, *Trends Polym. Sci.* **4**, 122 (1996).
55. A. Guyot, *Colloid Surface A: Physicochem. Eng. Aspects* **153**, 11 (1999).
56. I. Capek, *Adv. Polym. Sci.* **145**, 1 (1999).
57. H. Kawaguchi, *Prog. Polym. Sci.* **25**, 1171 (2000).
58. I. Capek, *Adv. Colloid Interface Sci.* **91**, 295 (2001).
59. I. Capek, *Adv. Colloid Interface Sci.* **99**, 77 (2002).
60. M. F. Cunningham, *Prog. Polym. Sci.* **27**, 1039 (2002).
61. M. S. El-Aasser and C. M. Miller, Preparation of latexes using miniemulsions, in *Polymeric Dispersions—Principles and Applications*, J. M. Asua (Ed.), Kluwer, Dordrecht, 1997, pp. 109–126.
62. E. D. Sudol and M. S. El-Aasser, Miniemulsion polymerization, in *Emulsion Polymerization and Emulsion Polymers*, P. A. Lovell and M. S. El-Aasser (Eds.), Wiley, Chichester, 1997, pp. 699–722.
63. M. Antonietti and K. Landfester, *Prog. Polym. Sci.* **27**, 689 (2002).
64. J. M. Asua, *Prog. Polym. Sci.* **27**, 1283 (2002).
65. C. S. Chern, Microemulsion Polymerization, in *Encyclopedia of Polymer Science and Technology*, Wiley-Interscience, New York, 2003.
66. J. W. Vanderhoff, H. L. Tarkowski, J. B. Schaffer, E. B. Bradford, and R. M. Wiley, *Adv. Chem. Ser.* **34**, 32 (1962).
67. F. Candau in *Polymer Association Structures: Microemulsions and Liquid Crystals*, M. El Nokaly (Ed.), ACS Symposium Series 384, Washington, D.C., 1989, p. 47.
68. F. Candau, in *Scientific Methods for the Study of Polymer Colloids and Their Applications*, F. Candau and R. H. Ottewill (Eds.), NATO ASI Series, D. Reidel, Dordrecht, p. 73.
69. B. V. Derjaguin and L. Landau, *Acta Physicochim.* **14**, 633 (1941).
70. W. B. Russel, D. A. Saville, and W. Schowalter, *Colloidal Dispersions*, Cambridge University Press, Cambridge, 1989.
71. J. W. Goodwin, *Colloids and Interfaces with Surfactants and Polymers—An Introduction*, John Wiley & Sons, West Sussex, 2004, Chapter 5.

72. A. Ya. Malkin and A. I. Isayev, *Rheology: Concepts, Methods, and Applications*, ChemTec Publishing, Toronto, 2006.

73. M. Mooney, *J. Colloid Sci.* **6**, 162 (1951).

74. Z. W. Wicks Jr., *Film Formation, Federation Series on Coatings Technology*, Federation of Societies for Coatings Technology, Philadelphia, 1986.

2

INTERFACIAL PHENOMENA

There is no doubt that the discipline of interfacial phenomena is an indispensable part of emulsion polymerization. Thus, the goal of this chapter is to offer the reader an introductory discussion on the interfacial phenomena related to the emulsion polymerization process, industrial emulsion polymerization processes (primarily the semibatch and continuous reaction systems), some important end-use properties of latex products, and some industrial applications. In this manner, the reader may effectively grasp the key features of emulsion polymerization mechanisms and kinetics. Some general readings in this vital interdisciplinary research area [1–6] are recommended for those who need to familiarize themselves with an introduction to the basic concepts of colloid and interface science.

2.1 THERMODYNAMIC CONSIDERATION

2.1.1 Emulsification of Oil in Water

In emulsion polymerization, the monomer-swollen polymer particle–water interfacial area is continuously generated during the particle nucleation and growth stages. What is the significance of the expanding total particle surface area with the progress of polymerization?

First, considering the dispersion of hydrophobic oil in the continuous aqueous phase to form an oil-in-water (O/W) emulsion, an exceedingly large oil–water interfacial area is created during the emulsification process (Figure

Principles and Applications of Emulsion Polymerization, by Chorng-Shyan Chern

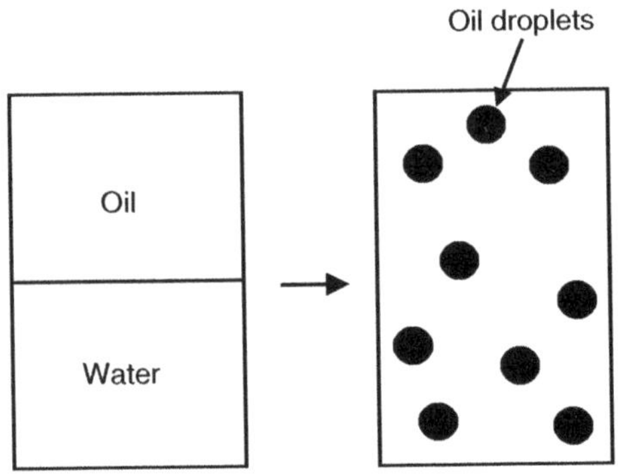

Figure 2.1. Emulsification of hydrophobic oil in water to form an oil-in-water emulsion.

2.1). It should be noted that the dimension of common emulsion droplets is considerably larger than the normal colloidal particle size range, which is ~1–10 μm in diameter. Thus, the change of the total oil–water interfacial area during the emulsification process ($\Delta A = A_f - A_i = A_f$) is approximately equal to the total droplet surface area at the final emulsification and prereaction state (A_f). This is simply because, for constant oil weight, the total droplet surface area is inversely proportional to the droplet size, and thus the total droplet surface area at the initial state (A_i) corresponding to the bulk oil phase is negligible compared to A_f. The corresponding Gibbs free energy change (ΔG) can be expressed as

$$\begin{aligned} \Delta G &= \Delta H - T\Delta S \\ &= \gamma \Delta A - T\Delta S \end{aligned} \tag{2.1}$$

where ΔH and ΔS are the enthalpy change and entropy change, respectively. The parameter γ is the oil–water interfacial tension, and T is the absolute temperature. The contribution of the second term on the right-hand side to ΔG is insignificant compared to the first term because the dimension of emulsified oil droplets is not very small (~10^3–10^4 nm in diameter). Thus, Eq. (2.1) can be further simplified to the following form:

$$\Delta G = \gamma \Delta A > 0 \tag{2.2}$$

This relationship indicates that emulsification is not a thermodynamically feasible process and significant work is required to homogenize the mixture of oil and water into minute oil droplets dispersed in water. In the absence of effective stabilization mechanisms, the resultant oil emulsion is extremely unstable due to the very high Gibbs free energy. This then promotes the coalescence of oil droplets and subsequent phase separation.

Incorporation of surface-active agents (surfactants) into emulsion formulations improves the compatibility between the oil droplets and the continuous aqueous phase. Surfactant molecules adsorbed on the droplets result in a reduction in the interfacial tension, lower the Gibbs free energy change during the emulsification process (Eq. (2.2)), and therefore enhance the emulsion stability. Nevertheless, the emulsion system is kinetically stable only because the Gibbs free energy change is still greater than zero. This example manifests the important concept that most common colloidal dispersions are thermodynamically unstable. Thus, the colloidal particles tend to flocculate with one another so that the total particle surface area is greatly reduced to alleviate the instability problem. Thus, the level of surfactants used in emulsion polymerization not only governs the total number of particles nucleated per unit volume of water but also regulates the size of the population of particles in the particle growth stage. For example, all the particle nuclei produced in the early stage of polymerization may survive the whole emulsion polymerization process, provided that a sufficient supply of surfactant is available for stabilizing the growing particles. On the other hand, severe particle flocculation may alter the size of the population of latex particles during polymerization if the concentration of surfactant is extremely low in the reaction medium.

2.1.2 Interfaces

As discussed in the previous section, the emulsification process involves the dispersion of one liquid (e.g., oil) in another liquid (e.g., water) and the interface between the disperse phase and continuous phase is of great importance. The specific surface area, defined as the total particle surface area per unit mass (A), can be calculated as follows:

$$A = 6/d\rho \tag{2.3}$$

where d and ρ are the diameter and density, respectively, of the particles. This relationship illustrates an important feature of colloidal dispersions. At constant weight of the constituent materials in the disperse phase, the total particle surface area is inversely proportional to the particle size. Thus, for example, the total particle surface area increases by seven orders of magnitude when the particle diameter is reduced from 1 m (bulk) to 100 nm (nano).

In the interior of a colloidal particle, the interactions between molecules act in a symmetric manner. By contrast, an imbalance in the intermolecular forces exists at the particle surface since the local environment changes significantly. This then results in a surface tension that tends to minimize the particle surface area. In addition, the fraction of molecules residing at the particle surface increases rapidly with decreasing particle size. It is no wonder, therefore, that many colloidal particles with a dimension in the nanometer range show quite unique performance properties in comparison with the bulk materials.

2.1.3 Surfactant Molecules Adsorbed at an Interface

As discussed above, surface-active agents (or surfactants) are generally required to stabilize the colloidal dispersions. Surfactant molecules show a strong tendency to be adsorbed onto the particle surfaces, reduce the particle–water interfacial tension, and thereby help to impart adequate stabilization to the colloidal system. The Gibbs adsorption equation that is often used to describe the concentration of surfactant adsorbed at the particle surfaces is:

$$\Gamma_{21} = -1/(RT)(d\sigma/d\ln c) \tag{2.4}$$

where Γ_{21} represents the surface excess of surfactant relative to the solvent, R the gas constant, T the absolute temperature, σ the particle–water interfacial tension, and c the concentration of surfactant in the continuous phase. It is very difficult to measure σ directly, but rather easy to determine the amount of adsorbed surfactant directly. Thus, Eq. (2.4) along with Γ_{21} data determined experimentally can be used to estimate the particle–water interfacial tension. Such information is of great importance in understanding why and how some molecules are adsorbed at an interface and how such adsorption has an influence on the characteristics of the interface.

2.2 SURFACTANTS

Surfactants are those molecules that comprise both hydrophilic and hydrophobic groups simultaneously. The hydrophilic groups can be either ionic (e.g., $-SO_4^-$, $-SO_3^-$, –COOH, and $-N(CH_3)_3^+$) or nonionic (e.g., $-O-(CH_2-CH_2-O)_n-H$). As to the hydrophobic groups, the most widely used ones are the alkyl chains ($-C_nH_{2n+1}$) and arakyl chains ($-C_nH_{2n+1}-C_6H_4-$). Representative surfactants include anionic sodium dodecyl sulfate (SDS), sodium dodecyl benzene sulfonate (SDBS), sodium bis(2-ethylhexyl) sulfosuccinate, sodium stearate, cationic hexadecyltrimethylammonium bromide, and nonionic *i*-octophenol polyethoxylate with an average of 40 monomeric units of ethylene oxide per molecule. Such amphoteric species tend to diffuse toward the interface between the oily and aqueous phases and reside therein.

The major function of surfactants is to stabilize the particle nuclei generated during the early stage of emulsion polymerization. Furthermore, the growing polymer particles also need to acquire surfactant molecules from those species that are adsorbed on the monomer droplets or dissolved in the continuous aqueous phase. Although the levels of surfactants used in typical emulsion polymerization recipes are relatively low (<5%), these surface-active agents play a crucial role in the particle nucleation and growth processes. It is therefore necessary for polymer scientists to understand the very basic concept about these surface-active agents.

2.2.1 Critical Micelle Concentration (CMC)

Perhaps the most fascinating behavior of many surfactant molecules in aqueous solution is the formation of micelles. The major interactions between surfactant and water include (a) the interactions between the hydrophobic group of surfactant and water, (b) the attractive force between the hydrophobic groups belonging to adjacent surfactant molecules, (c) the solvation of the hydrophilic group of surfactant by water, (d) the interactions between solvated hydrophilic groups belonging to adjacent surfactant molecules, (e) the interactions between the ionic group of surfactant and co-ion, and (f) the geometric and packing constrains originating from the particular molecular structure of surfactant. At relatively low surfactant concentration, surfactant molecules can be dissolved in the aqueous phase and adsorbed on the water–air interface or the interior wall of the container. As the surfactant concentration is increased, the surfactant molecules adsorbed at the water–air interface become more crowded. Immediately after a critical surfactant concentration is reached, the water–air interface becomes saturated with the adsorbed surfactant molecules and the aggregated surfactant molecules (i.e., micelles comprising ~10^1–10^2 surfactant molecules per micelle) start to form in the continuous aqueous phase as a result of the complex interactions mentioned above. This surfactant concentration is defined as the critical micelle concentration (CMC). In the emulsion polymerization of relatively hydrophobic monomers such as styrene and butadiene, the monomer-swollen micelles are considered as the primary loci for particle nucleation [7–12].

Measurements of the bulk solution properties (e.g., surface tension, electrical conductivity, fluorescence, and light scattering intensity) as a function of surfactant concentration can be used to determine the CMC. As shown schematically in Figure 2.2, the point at which the sudden change in surface tension occurs is taken as the CMC of the aqueous surfactant solutions. How to establish the correlation between the CMC data and the molecular structure of surfactants is of great importance in the selection of optimum surfactants for the effective stabilization of various emulsion polymerization systems. This subject will be the focus of the following discussion.

2.2.2 Hydrophile–Lipophile Balance (HLB)

The concept of hydrophile–lipophile balance (HLB) was first developed by Griffin [13] to correlate the structure of surfactant molecules with their surface activity. The HLB number (0–20) reflects the hydrophilicity of surfactant, and it increases with increasing hydrophilicity. A general trend often observed in a family of surfactants is the increased CMC with HLB. Table 2.1 serves as a general guide for the formulator to choose surfactants that are most suited to meet the requirements of end-users. This semiempirical approach has been proved to be quite useful. Davies and Rideal [14] proposed that the HLB value of a particular surfactant could be calculated according to the group contribution approach.

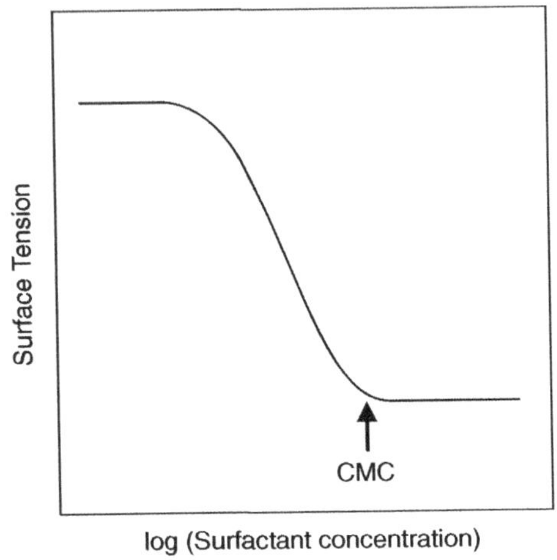

Figure 2.2. Surface tension as a function of surfactant concentration. The point at which the sudden change in surface tension occurs is taken as the CMC of the aqueous surfactant solutions, as indicated by the arrow.

Table 2.1. Ranges of HLB Values and Corresponding Areas of Applications

HLB Range	Application
3–6	Water-in-oil emulsions
7–9	Wetting
8–18+	Oil-in-water emulsions
3–15	Detergency
15–18	Solubilization

$$\text{HLB} = 7 + \sum(\text{hydrophilic group number}) - \sum(\text{lipophilic group number}) \tag{2.5}$$

Table 2.2 lists the group numbers for various chemical groups, which were taken from the literature [14]. Taking sodium dodecyl sulfate (SDS) as an example, its HLB value is estimated to be $7 + 38.7 - [(1 \times 0.475) + (11 \times 0.475)] = 40.0$.

The use of surfactant mixtures (e.g., an anionic surfactant in combination with a nonionic surfactant) often results in more stable colloidal dispersions due to the synergistic effect. As a first approximation, the HLB number of a surfactant mixture (HLB_{mix}) can be estimated by the following equation:

$$\text{HLB}_{\text{mix}} = \sum f_i \text{HLB}_i \tag{2.6}$$

where f_i and HLB_i are the weight fraction and HLB number of surfactant i in the mixture, respectively, and $\Sigma f_i = 1$. It should be noted that Eq. (2.6) should

Table 2.2. Group Numbers for a Variety of Chemical Groups Used to Calculate the HLB Values of Surfactants [14]

Type	Chemical Group	Group Number
Hydrophilic	$-SO_4Na$	38.7
	$-COOK$	21.1
	$-COONa$	19.1
	$-SO_3Na$	11.0
	$=N-$	9.4
	Ester (sorbitan ring)	6.8
	Ester (free)	2.4
	$-COOH$	2.1
	$-OH$ (free)	1.9
	$-O-$	1.3
	$-OH$ (sorbitan ring)	0.5
Hydrophobic	$-CH-$	0.475
	$=CH-$	0.475
	$-CH_2-$	0.475
	$-CH_3$	0.475

be regarded as qualitative only. Nevertheless, this empirical approach still serves as a powerful tool for surfactant formulation purposes.

2.2.3 Solubility Parameter

Another important parameter that characterizes the intermolecular interactions of materials and verifies the HLB approach in a theoretical sense is the solubility parameter (δ). This theoretical approach pioneered by Hildebrand is primarily based on the overall attractive forces between neighboring molecules in a condensed phase. He defined the solubility parameter to characterize the cohesive energy density of a material. The higher the cohesive energy density of a material, the greater the propensity for adjacent molecules to tightly attract one another. For example, water (molecular weight (MW) = 18) has a cohesive energy density of $\delta = 47.9\ (\mathrm{J\,cm^{-3}})^{1/2}$, a boiling point ($T_b$) of 100 °C and a viscosity (η) of 1 cP at 20 °C, whereas the values of δ, T_b, and η for diethyl ether (MW = 74) are 15.1 $(\mathrm{J\,cm^{-3}})^{1/2}$, 35 °C, and 0.23 cP at 20 °C, respectively. The differences in these physical properties between water and diethyl ether are primarily attributed to the different intermolecular interactions (i.e., hydrogen bonding, dipole–dipole, and van der Waals interactions for water versus dipole–dipole and van der Waals interactions for diethyl ether). As expected, water and diethyl ether have very different values of δ, and therefore they are immiscible.

The basic idea behind the use of δ in identifying the optimum surfactant for stabilizing a particular colloidal system is best illustrated by the following

Table 2.3. Solubility Parameter [$(J\,cm^{-3})^{1/2}$] Data for Some Representative Materials [15]

Material	δ	Material	δ	Material	δ
Alkanes		**Alcohols**		**Nitrogen-Containing Compounds *(continued)***	
n-Heptane	15.3	Ethanol	26.5	Ethanol amine	31.5
n-Decane	15.8	1-Propanol	24.5	Pyridine	21.8
n-Hexadecane	16.4	1-Butanol	23.1	Diethyl amine	16.8
Cyclohexane	16.8	Benzyl alcohol	23.8	Formamide	36.6
Halogenated Compounds		1-Dodecanol	20.0	**Miscellaneous**	
		Glycerol	33.7	Dimethyl sulfide	19.2
Chloroform	19.0	**Ethers**		Dimethyl carbonate	20.3
Trichloro ethylene	19.0	Diethyl ether	15.8	Thiophene	20.1
Chlorobenzene	19.6	Dibenzyl ether	19.3	Tricresyl phosphate	17.2
Aldehydes and Ketones		Methoxy benzene	19.5	Triphenyl phosphate	17.6
		Aromatics		Dimethylsulfoxide	24.5
Acetaldehyde	21.1	Benzene	18.6	Dimethyl siloxanes	10–12
Benzaldehyde	21.5	Toluene	18.2	Water	47.9
2-Butanone	19.0	Naphthalene	20.3		
Acetophenone	21.8	Styrene	19.0		
Furfural	24.4	**Nitrogen-Containing Compounds**			
Esters					
Ethyl acetate	18.1	Acetonitrile	24.4		
n-Butyl acetate	17.4	Benzonitrile	19.9		
Di-*n*-butylphthalate	20.2	Nitrobenzene	22.2		

example. Which surfactant is most effective in emulsifying styrene ($\delta = 19.0$ $(J\,cm^{-3})^{1/2}$) in water ($\delta = 47.9$ $(J\,cm^{-3})^{1/2}$)? As would be expected, the surfactant of choice must have a hydrophobic group with a value of δ close to 18.2 $(J\,cm^{-3})^{1/2}$ in combination with a hydrophilic group with a value of δ close to 47.9 $(J\,cm^{-3})^{1/2}$. As a rule of thumb, excellent compatibility (i.e., a very high degree of interaction) between two materials can be achieved if the difference in δ values is within ±4 $(J\,cm^{-3})^{1/2}$. Any surfactant with both the hydrophobic and hydrophilic groups satisfying this criterion will perform reasonably well in this respect.

The δ data for some representative materials are collected in Table 2.3 [15]. The hydrophilicity of compounds increases with increasing δ. The δ and HLB data for some simple surfactants are summarized in Table 2.4 [15]. What if the δ data for some materials are not available in the literature? Just like the HLB approach, the equation based on the group contribution method for calculating δ at 25 °C is shown below [16].

$$\delta_d = \sum F_g / V \tag{2.7}$$

where the subscript d on δ indicates that the calculation is based on the dispersion component of the cohesive energy density, F_g is the group contribution

Table 2.4. Solubility Parameter and HLB Data for Some Simple Surfactants [15]

Surfactant	δ [(J cm^{-3})$^{1/2}$]	HLB
Oleic acid	16.8	1.0
Glycerol monostearate	17.0	3.8
Sorbitan monolaurate	17.6	8.6
Polyoxyethylene (10) oleyl ether	18.2	12.4
Polyoxyethylene (10) cetyl ether	18.6	15.7
Sodium octadecanoate (stearate)	19.0	18.0
Sodium hexadecanoate (palmitate)	19.2	19.0
Sodium dodecanoate (laurate)	19.6	20.9
Sodium dinonylnaphthalene sulfonate	21.5	28.5
Sodium dioctyl sulfosuccinate	25.0	32.0
Sodium dodecyl sulfate	28.8	40.0
Sodium decyl sulfate	30.0	40.0
Sodium octyl sulfate	32.3	41.9

Table 2.5. Some Molar Attraction Constants (Group Numbers) for Calculation of Solubility Parameter at 25 °C[16]

Group	F_g	Group	F_g
$-CH_3$	438	Conjugation	40–60
$-CH_2-$ (single bonded)	272	–H (variable)	160–205
$-CH<$	57	–O– (ether)	143
$>C<$	190	$>C=O$ (ketones)	563
$CH_2=$	389	–C(O)–O– (esters)	634
–CH= (double bonded)	227	$-C\equiv N$	839
$>C=$	39	–Cl (single)	552
$HC\equiv C-$	583	–Cl ($>CCl_2$)	532
$-C\equiv C-$	454	–Cl ($>CCl_3$)	511
Phenyl	1503	–Cl ($>C(Cl)-C(Cl)<$)	497
Phenylene (*o*, *m*, *p*)	1346	–Br (single)	340
Naphthyl	2344	–I (single)	425
5-Membered ring	215–235	$-CF_2$ (*n*-fluorocarbons)	150
6-Membered ring	195–215	$-CF_3$ (*n*-fluorocarbons)	274
–OH (single)	348	$-PO_4$ (organic phosphate)	1020
–S– (sulfides)	460	$-ONO_2$ (nitrate)	900
–SH (thiol)	644	$-NO_2$ (aliphatic nitro)	900

for each simple unit, and V is the molecular volume. It should be noted that a reasonable estimation for polar and even hydrogen-bonding materials can be achieved by Eq. (2.7). Some molar attraction constants (i.e., group numbers) for the calculation of δ at 25 °C are listed in Table 2.5 [16]. For example, for one mole of the monomeric units of polystyrene ($-(CH_2C(H)C_6H_5)_n-$, MW = 104 g mol^{-1}, ρ = 1.05 g cm^{-3}), the value of δ is estimated to be (272 + 57 +

1503)/(104/1.05) = 18.5 $(\mathrm{J\,cm^{-3}})^{1/2}$, which is quite close to the value of δ for the styrene monomer listed in Table 2.3.

Considering the emulsification of oil in water, the strategy is to match the hydrophobic and hydrophilic groups of surfactant, respectively, with the oily and continuous aqueous phases. This can be achieved by manipulating the molecular structure of surfactant or by adjusting the composition of one or both phases. When the choice of surfactant is limited, one can adopt the concept of solubility parameter to effectively modify the recipe. As a first approximation, the solubility parameter of a mixture (δ_{mix}) can be estimated by the following equation:

$$\delta_{\text{mix}} = \sum x_i \delta_i \tag{2.8}$$

where x_i and δ_i are the mole fraction and solubility parameter of component i in the mixture, respectively, and $\Sigma x_i = 1$.

2.3 COLLOIDAL STABILITY

Colloidal stability is the resultant of various interparticle interactions such as the attractive van der Waals forces and the repulsive electrostatic and steric interactions [17–21]. In the absence of adequate stabilization mechanisms, the colloidal dispersion starts to lose its stability upon aging. Accompanied with particle flocculation, the Gibbs free energy of the colloidal system decreases continuously at the expense of the total particle surface area. This is undesirable because acceptable colloidal stability (e.g., six months of storage stability required for most common latex products) is a prerequisite for practical industrial applications. Flocculation of particle nuclei or growing particles during emulsion polymerization also has a significant influence on the resultant latex particle size and particle size distribution and the level of scraps adhered to the reactor wall and agitator and filterable solids. Unfortunately, most scientists working on emulsion polymers do not have an adequate training in this interdisciplinary field. The objective of this section was therefore to provide the reader with introductory information on the origin of various fundamental interparticle interactions.

2.3.1 Van der Waals Forces

The Gibbs free energy of the universal van der Waals interactions (ΔG_v) between two identical spheres with radius a separated by a distance H in a vacuum ($H/a \ll 1$) can be calculated by the following equation:

$$\Delta G_v = -A_H a/(12\mathrm{H})[1 + 3/4(H/a)] \tag{2.9}$$

where A_H is the Hamaker constant. The negative sign for ΔG_v indicates that the van der Waals interaction is attractive in nature. The higher terms involving

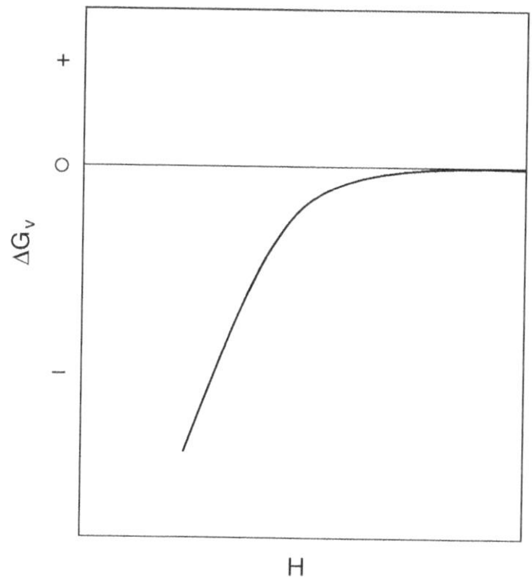

Figure 2.3. A schematic representation of the van der Waals interaction between two particles.

$(H/a)^2$, $(H/a)^3$, and so on, inside the bracket of Eq. (2.9) can be safely neglected because the ratio H/a is much smaller than unity. A schematic representation of the van der Waals interaction between two approaching particles is shown in Figure 2.3. It is the exceedingly deep potential energy well observed at small distance of separation between two interactive particles that is responsible for the loss of colloidal stability. The attractive van der Waals force decreases rapidly with increasing distance of separation.

For the colloidal particles dispersed in some medium other than a vacuum, the attractive van der Waals forces between two particles will be reduced due to the effect of the molecules present in the medium. To derive an expression for the effective Hamaker constant ($A_{H,\text{eff}}$) for the particles (component 2, •) dispersed in a medium (component 1, ○), consider the following change of the van der Waals interactions:

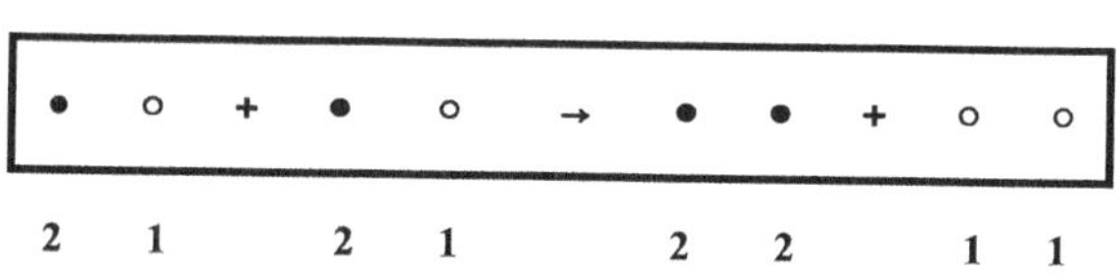

where the symbols • ○, • •, and ○ ○ represent the van der Waals interactions between the components 2 and 1, between the components 2 and 2, and between the components 1 and 1 in a vacuum, respectively. The Gibbs free energy change for the above process can be written as

$$\Delta G_{\text{eff}} = G_{11} + G_{22} - 2G_{12} \tag{2.10}$$

With the assumption that the Gibbs free energy terms are linearly proportional to the corresponding Hamaker constants, then Eq. (2.10) can be further reduced to the following relationships:

$$\begin{aligned} A_{H,\text{eff}} &= A_{H,11} + A_{H,22} - 2A_{H,12} \\ &\fallingdotseq A_{H,11} + A_{H,22} - 2(A_{H,11}A_{H,22})^{1/2} \\ &= (A_{H,11}{}^{1/2} - A_{H,22}{}^{1/2})^2 \end{aligned} \tag{2.11}$$

where $A_{H,11}$ and $A_{H,22}$ represent the Hamaker constants for the components 1 and 2 in a vacuum, respectively, and the value of $A_{H,12}$ is further assumed to be equal to the geometric average of the Hamaker constants for the component 1 and 2 in a vacuum. The effective Hamaker constant should be used in Eq. (2.9) to calculate the attractive van der Waals forces between the particles in a particular medium. It is also interesting to note that the van der Waals interaction is negligible as the difference between $A_{H,11}$ and $A_{H,22}$ approaches zero.

Table 2.6 lists some representative Hamaker constants (A_H) for the materials in a vacuum. For demonstration, the Hamaker constants for polystyrene and water in a vacuum are 6.57×10^{-20} and 4.35×10^{-20} J, respectively. Thus, the effective Hamaker constant for the polystyrene particles dispersed in water is $[(4.35 \times 10^{-20})^{1/2} - (6.57 \times 10^{-20})^{1/2}]^2 = 2.28 \times 10^{-21}$ J.

In a similar manner, the effective Hamaker constant for the different particles (components 2 and 3) dispersed in the medium (component 1) can be estimated by the following relationships:

Table 2.6. Some Representative Hamaker Constants for the Materials in a Vacuum

Substance	$A_H \times 10^{20}$ (J)	Reference	Substance	$A_H \times 10^{20}$ (J)	Reference
Water	0.6–0.7	22	Mica	9.7	27
	4	23		10	25
	3.70	24	Gold	20–40	23
	3.7–4.0	25		45.3	1
	4.35	1	*n*-Octane	4.26–5.02	26
Polystyrene	6.37–6.58	24		5.02	1
	6.57	1		4.5	25
	6.6–7.9	25	*n*-Hexadecane	6.2–6.31	26
	7.8–9.8	26		6.31	1
Polytetrafluoroethylene	2.75–3.8	24		5.2	25
	3.8	25	Benzene	5.0	25
Silica	6.55	24			
	6.5	25			

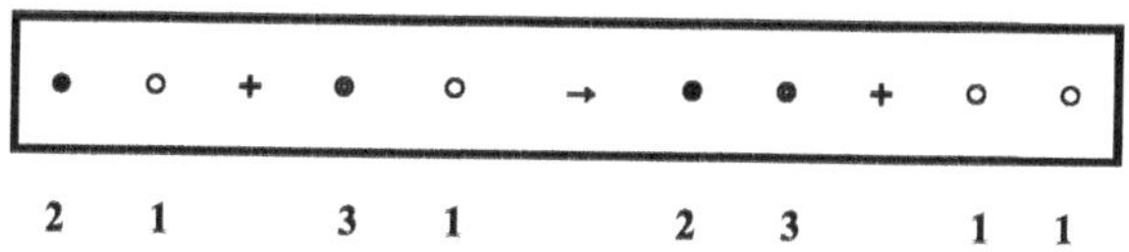

$$\Delta G_{\text{eff}} = \Delta G_{11} + \Delta G_{23} - \Delta G_{12} - \Delta G_{13} \tag{2.12}$$

$$\begin{aligned} A_{H,\text{eff}} &= A_{H,11} + A_{H,23} - A_{H,12} - A_{H,13} \\ &\fallingdotseq A_{H,11} + (A_{H,22}A_{H,33})^{1/2} - (A_{H,11}A_{H,22})^{1/2} - (A_{H,11}A_{H,33})^{1/2} \\ &= (A_{H,22}{}^{1/2} - A_{H,11}{}^{1/2})(A_{H,33}{}^{1/2} - A_{H,11}{}^{1/2}) \end{aligned} \tag{2.13}$$

where $A_{H,33}$ is the Hamaker constant for the component 3 in a vacuum. Equation (2.13) is useful in the colloidal stability study of composite particles (e.g., compounding of emulsion polymers and inorganic particles). For example, the Hamaker constants for polystyrene, silica, and water in a vacuum are 6.57×10^{-20}, 6.5×10^{-20}, and 4.35×10^{-20} J, respectively. Thus, the effective Hamaker constant for the interaction of polystyrene particles and the silica particles dispersed in water are 2.28×10^{-21} and 2.15×10^{-21} J, respectively. Furthermore, the effective Hamaker constant for the polystyrene and silica pair dispersed in water is $((6.57 \times 10^{-20})^{1/2} - (4.35 \times 10^{-20})^{1/2})\ ((6.5 \times 10^{-20})^{1/2} - (4.35 \times 10^{-20})^{1/2}) = 2.21 \times 10^{-21}$ J. The attractive van der Waals interaction in decreasing order is the polystyrene–polystyrene pair > the polystyrene–silica pair > the silica – silica pair. Considering another colloidal system comprising polystyrene and gold (Hamaker constant in a vacuum = 4.53×10^{-19} J) particles dispersed in water, the effective Hamaker constants for the polystyrene–polystyrene pair, the gold–gold pair, and the polystyrene–gold pair in water are 2.28×10^{-21}, 2.16×10^{-19} and 2.22×10^{-20} J, respectively. In this case, the attractive van der Waals interaction in decreasing order is the gold–gold pair >> the polystyrene–gold pair >> the polystyrene–polystyrene pair. The larger the effective Hamaker constant, the stronger the tendency for the pair of colloidal particles to coagulate with each other. Apparently, the former example is more stable than the latter. The calculation of effective Hamaker constants may thus disclose the potential instability problem associated with a particular colloidal dispersion if present.

Some effective mechanisms must be incorporated into a colloidal system in order to impart a high enough potential energy barrier to stabilize the interactive particles against the van der Waals attraction forces that pull them together. For hydrophobic particles dispersed in the continuous aqueous phase, two primary mechanisms including the electrostatic repulsion force between two similarly charged particles and the steric (or entropic) repulsion force between two particles coated with a layer of hydrophilic polymer chains are used to stabilize the colloidal dispersions. The following section deals with the basic principle (DLVO theory) behind the electrostatic stabilization mechanism. This is followed by a brief discussion on the steric stabilization mechanism and then the flocculation kinetics to conclude this chapter.

2.3.2 Electrostatic Interactions

For typical aqueous colloidal dispersions, the particles may carry some charges most likely due to the preferential (or differential) dissolution of particle surface ions, direct ionization of particle surface groups, substitution of particle surface ions, specific adsorption of ions, and particle surface charges originating from specific crystal structures.

For example, silver iodide (AgI) has a very limited solubility in water, and its solubility product is given by $K_{sp} = [Ag^+][I^-] = 10^{-16}$ M, where $[Ag^+]$ and $[I^-]$ are the concentrations of Ag^+ and I^-, respectively. When the water-soluble $AgNO_3$ is added to the AgI colloidal dispersion, dissolution of Ag^+ from the AgI particle surface is greatly retarded due to the common ion effect. As a result, the particle surface I^- concentration (i.e., net negative charge) decreases with increasing $[Ag^+]$. At some characteristic $[Ag^+]$, the net AgI particle surface charge is equal to zero (termed the point of zero charge). Further increasing $[Ag^+]$, the net particle surface charge becomes positive.

Another mechanism that allows the hydrophobic colloidal uolume fraction of partictes charges is the specific adsorption of surface-active ions (e.g., $C_{12}H_{25}SO_4^-$). This phenomenon has a significant influence on the particle nucleation and growth stages in emulsion polymerization. The major role of the adsorbed anionic surfactant molecules is to impart the electrostatic stabilization effect to the newly born and growing particles during polymerization. The colloidal system is considered to be electrically neutral. This implies that the amount of the particle surface charges are balanced by an equal amount of positive charges associated with the counterions in the continuous aqueous phase. The distribution of counterions close to the negatively charged particles is not uniform. The probability of finding counterions decreases continuously from the particle surface to the bulk aqueous solution. On the other hand, the concentration of co-ions increases with increasing distance from the particle surface, as shown in Figure 2.4. A layer of nonuniformly distributed ions surrounding the particle is defined as the electrical double layer, which is responsible for the electrostatic stabilization effect [17–19]. A schematic model for the structure of the electrical double layer is shown in Figure 2.5.

The pioneering Gouy–Chapman theory can be used to quantitatively describe the diffuse electrical double layer. The electrical potential at a distance x from the colloidal particle surface ($\psi(x)$) is described by the one-dimensional Poisson's equation that relates the number of charges per unit volume (or space charge density, ρ) to $\psi(x)$.

$$d^2\psi(x)/dx^2 = -\rho/\varepsilon_0\varepsilon_r \tag{2.14}$$

where ε_0 and ε_r are the dielectric constant of a vacuum and the relative dielectric constant of the medium, respectively. The parameter ρ can be calculated by the following equation:

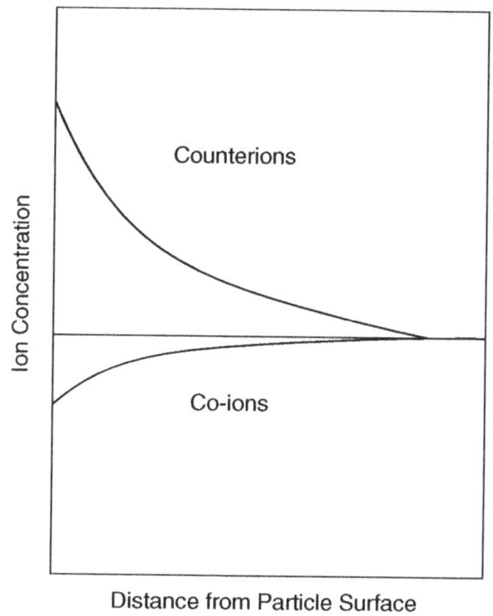

Figure 2.4. A schematic representation of the concentration profiles of counterions and co-ions in the diffuse electrical double layer.

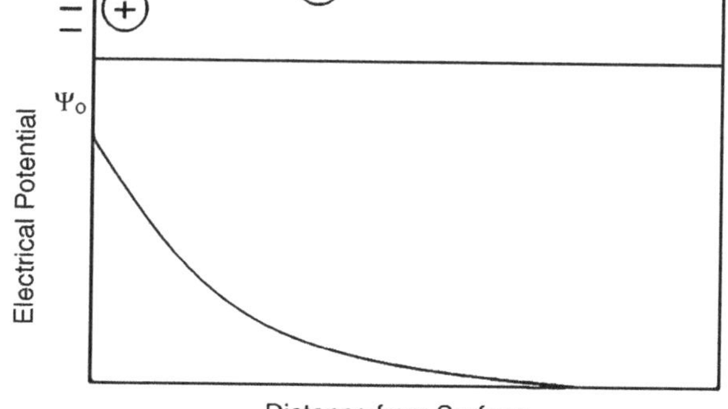

Figure 2.5. A schematic model for the structure of the diffuse electrical double layer surrounding the colloidal particle.

$$\rho = \sum n_i z_i e \tag{2.15}$$

where n_i and z_i are the number and valency of ion i, respectively, and e is the charge of an electron. Assuming that the distribution of ions within the electrical double layer follows the Boltzmann distribution (i.e., $n_i = n_{i0}\exp(-z_i e\psi(x)/(kT))$, k is Boltzmann constant, and T is absolute temperature) and a symmetric electrolyte is used, then Eq. (2.15) can be reduced to

$$\rho = -2n_0 ze \sinh[ze\psi(x)/(kT)] \tag{2.16}$$

Substituting Eq. (2.16) into Eq. (2.14) leads to the following second-order ordinary differential equation along with the two boundary conditions.

$$\begin{aligned} &d^2\psi(x)/dx^2 = [2n_0 ze/(\varepsilon_0\varepsilon_r)]\sinh[ze\psi(x)/(kT)] \\ &x = 0, \qquad \psi(x) = \psi_0 \text{ (surface electrical potential)} \\ &x = \infty, \qquad \psi(x) = 0, \qquad \text{and} \qquad d\psi(x)/dx = 0 \end{aligned} \tag{2.17}$$

Equation (2.17) can be integrated to give

$$\begin{aligned} \psi(x) = 2kT/(ze)\ln\{&[1+\exp(-\kappa x)\tanh(ze\psi_0/4kT)]/ \\ &[1-\exp(-\kappa x)\tanh(ze\psi_0/4kT)]\} \end{aligned} \tag{2.18}$$

where the reciprocal of $\kappa\left[=(\Sigma e z_i^2 n_0/\varepsilon_0\varepsilon_r kT)^{1/2}\right]$ is defined as the thickness of the diffuse electrical double layer (or Debye–Hückel length). For the condition that ψ_0 is smaller than 25 mV and x is small, Eq. (2.18) is further reduced to the well-known Debye–Hückel equation, as follows:

$$\psi(x) = \psi_0 \exp(-\kappa x) \tag{2.19}$$

The general features of the structure of the electrical double layer are briefly described as follows. First, the electrical potential decreases rapidly (exponentially) with increasing distance from the particle surface (Figure 2.6). At constant κ and x, the higher the surface potential, the higher the electrical potential (Figure 2.6a). Furthermore, the electrical double layer related parameter, κ, plays an important role in determining the electrical potential profiles. Increasing the concentration of electrolytes or the valency of counterions results in an increase in κ, thereby leading to a reduction in $\psi(x)$. As a result, the colloidal stability decreases significantly with increasing κ. This phenomenon is often referred to as the compression of the diffuse electrical double layer (Figure 2.6b,c), and this phenomenon has been widely used in the treatment of wastewater.

It should be noted that Eq. (2.14)–(2.19) were derived based on a flat plate immersed in the continuous aqueous phase. For a single spherical particle of

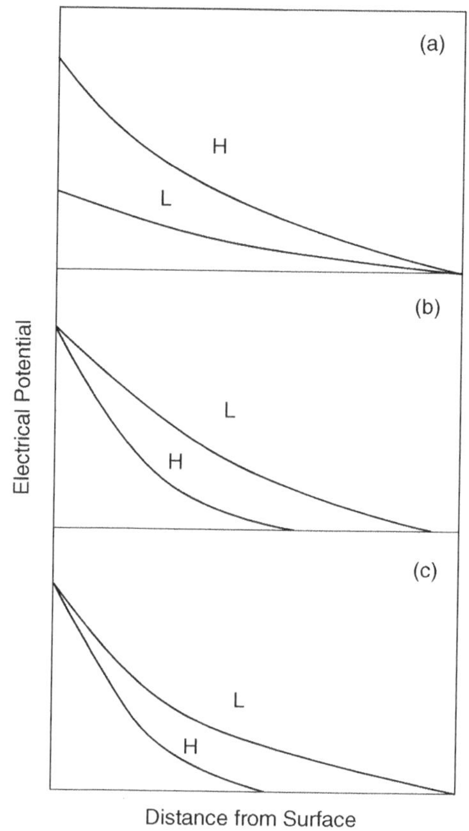

Figure 2.6. A schematic representation for the effects of **(a)** surface electrical potential, **(b)** concentration of electrolytes, and **(c)** valency of counterions on the electrical potential profiles. The symbols H and L represent the high and low levels of a particular parameter (e.g., surface electrical potential in Figure 2.6a), respectively.

a radius of a in the aqueous solution of a symmetric electrolyte, the corresponding Debye–Hückel equation in spherical polar coordinates becomes

$$\psi(r) = \psi_0(a/r)\exp[-\kappa(r-a)] \tag{2.20}$$

where r is the distance from the center of the particle.

For an isolated pair of the spherical particles, the Gibbs free energy of the repulsive electrostatic interactions (ΔG_e) can be calculated by the following equations.

(a) **Identical Spherical Particles at $\kappa a < 5$.** In this case, the thickness of the electrical double layer is of a similar magnitude to the particle radius.

With the Debye–Hückel approximation and relatively weak overlap of the diffuse electrical double layers, ΔG_e can be expressed as

$$\Delta G_e = 2\pi\varepsilon_0\varepsilon_r a\psi_0^2 \exp(-\kappa H) \quad (2.21)$$

where $H = r - 2a$ is the closest distance between the surfaces of the particles and r is the center-to-center distance. When the product of κ and a approaches 1 or less, Eq. (2.21) can be used to calculate ΔG_e for the two particles with a short distance of separation [28].

(b) **Identical Spherical Particles at $\kappa a > 10$.** This condition simply represents a relatively thin electrical double layer compared to the particle radius. For low surface potentials, ΔG_e can be estimated by the following equation:

$$\Delta G_e = 2\pi\varepsilon_0\varepsilon_r a\psi_0^2 \ln[1+\exp(-\kappa H)] \quad (2.22)$$

This equation is still acceptable for the colloidal system subject to the conditions of $\kappa a > 2$ and $\kappa H < 2^3$.

(c) **Spherical Particles of Radii a_1 and a_2 at $\kappa a > 10$.** The extended form of Eq. (2.22) that accounts for particles with different sizes was developed by Hogg et al. [29], as shown below.

$$\Delta G_e = \pi\varepsilon_0\varepsilon_r a_1 a_2 (\psi_{01}^2+\psi_{02}^2)/(a_1+a_2)\{2\psi_{01}\psi_{02}/(\psi_{01}^2+\psi_{02}^2)\times \ln[(1+\exp(-\kappa H))/(1-\exp(-\kappa H))] + \ln[(1-\exp(-2\kappa H))]\} \quad (2.23)$$

where the subscripts 1 and 2 correspond to the particles 1 and 2, respectively.

Ignoring the contribution of the steric stabilization effect at this point, the total Gibbs free energy of interactions between an isolated pair of particles (ΔG) can be expressed as the sum of the Gibbs free energy of the van der Waals interaction [Eq. (2.9)] and the Gibbs free energy of the electrostatic interaction [e.g., Eq. (2.21)].

$$\begin{aligned}\Delta G &= \Delta G_v + \Delta G_e \\ &= -A_H a/(12H)\left[1+\frac{3}{4}(H/a)\right] + 2\pi\varepsilon_0\varepsilon_r a\psi_0^2 \exp[-H/(1/\kappa)]\end{aligned} \quad (2.24)$$

The negative and positive signs of ΔG represent the net attractive and repulsive interactions between the particles, respectively. This theoretical framework that represents the first comprehensive study for the stability of colloidal dispersions has been termed the well-known DLVO theory [17, 18].

Equation (2.24) signifies the fact that the colloidal stability is primarily governed by the competition between the two opposite interparticle interactions that show quite different dependence on the closest distance between the surfaces of the particles: $\Delta G_v \sim 1/\mathrm{H}$ and $\Delta G_e \sim \exp(-H)$. This shows the most important characteristic of the DLVO theory, which states that ΔG_e decreases much faster with increasing H in comparison with ΔG_v. This is the reason why kinetically stable colloidal dispersions exist in the world. Furthermore, ΔG_v is primarily determined by the effective Hamaker constant (A_H), whereas ΔG_e is governed by the surface electrical potential (ψ_0) and the thickness of the diffuse electrical double layer ($1/\kappa$). The larger the A_H or the smaller the ψ_0, the worse the colloidal stability. As mentioned above, $\kappa \sim (\Sigma z_i^2 n_0)^{1/2}$; therefore, increasing the valency or concentration of counterions results in a rapid reduction in ΔG_e. Figure 2.7 illustrates the key features of the DLVO theory. For a specific family of colloidal dispersions (e.g., emulsion polymers), the effective Hamaker constants are relatively similar and, consequently, the surface electrical potential and thickness of the diffuse electrical double layer are the two major parameters that can be used to manipulate the colloidal stability.

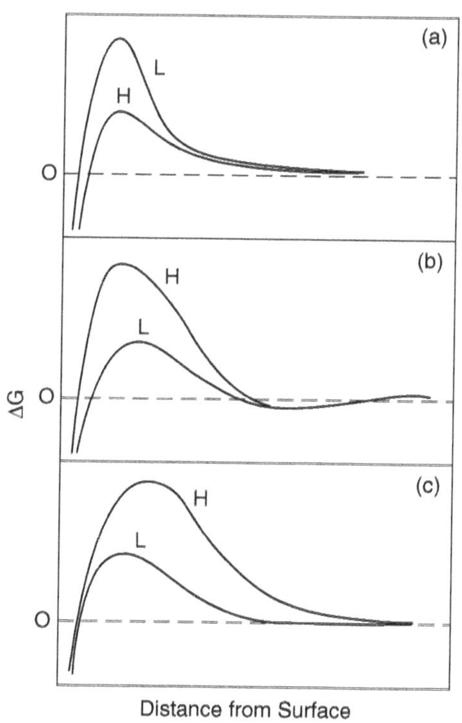

Figure 2.7. A schematic representation for the effects of **(a)** effective Hamaker constant, **(b)** surface electrical potential, and **(c)** thickness of the electrical double layer on the electrical potential profiles. The symbols H and L represent the high and low levels of A_H, ψ_0, or κ^{-1}, respectively.

The total potential energy barrier against particle flocculation decreases with increasing concentration of counterions, as would be predicted by Eq. (2.24). As a rule of thumb, the critical coagulation concentration (CCC) is achieved when the conditions of $\Delta G = 0$ and $d\,\Delta G/dH = 0$ are achieved.

$$\text{CCC} \sim A_H^{-2} z^{-6} \tag{2.25}$$

This is consistent with the empirical Schultze–Hardy rule [2, 30]. Based on this relationship, experiments can be designed to evaluate the effectiveness of colloidal dispersions to resist the added electrolytes.

It should be noted that the above discussion only dealt with the electrostatic interaction between an isolated pair of particles. It is reasonable to neglect the effect of the particle concentration for the dilute colloidal dispersion because the distance of separation between the particles is much larger than the thickness of the electrical double layer. Nevertheless, the counterions associated with other particles (macroions) have an appreciable influence on the pair of interactive particles when the particle concentration is so high that the distance of separation between the particles becomes comparable to the thickness of the diffuse electrical double layer. The following equation, which takes into consideration the effect of the particle concentration, was proposed to calculate κ [2]:

$$\kappa = \{[e^2/(\varepsilon_0\varepsilon_1 kT)][2z^2 n_0 - (3\sigma_0 z\varphi/(ae))]/(1-\varphi)\}^{1/2} \tag{2.26}$$

where σ_0 is the particle surface charge density and φ is the volume fraction of particles. The term $(1 - \varphi)$ appearing in the denominator of Eq. (2.26) represents the correction of the counterion concentration for the volume occupied by the charged particles. Figure 2.8 illustrates the effect of the particle concentration on the thickness of the electrical double layer. The thickness of the electrical double layer decreases (or κ increases) with increasing volume fraction of the dispersed phase. This implies that the colloidal stability decreases with increasing volume fraction of the dispersed phase due to the crowding effect. Figure 2.9 shows the effect of the particle concentration on the Gibbs free energy of the electrostatic interaction (ΔG_v). Thus, the particle concentration effect is a critical issue that needs to be taken into consideration in the preparation of colloidal dispersions with high solids contents.

Surfactant-free emulsion polymerization is an important industrial process for the manufacture of water-based polymeric materials with excellent water resistance and adhesion properties. In the absence of surfactant, the colloidal stability of the polymerization system can be achieved by using a persulfate initiator such as potassium persulfate. In addition to initiating free radical polymerization, the negatively charged sulfate end-groups ($-SO_4^-$) of oligomeric radicals originating from the persulfate initiator molecules, which are anchored on the surfaces of particle nuclei, act as an electrostatic stabilizer that impart colloidal stability to the latex particles. Due to the relatively low particle

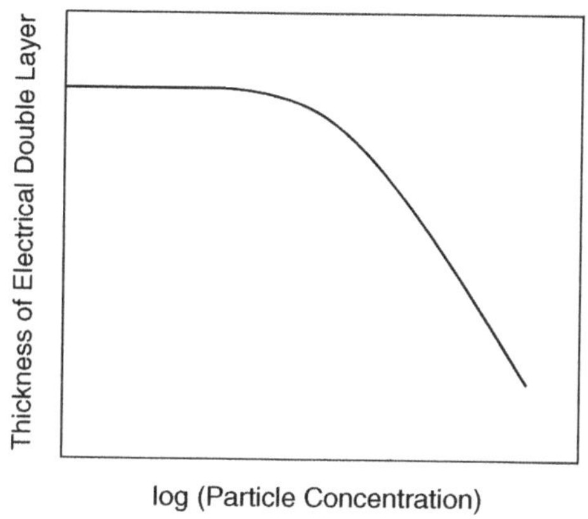

Figure 2.8. A schematic representation for the effect of the particle concentration on the thickness of the diffuse electrical double layer.

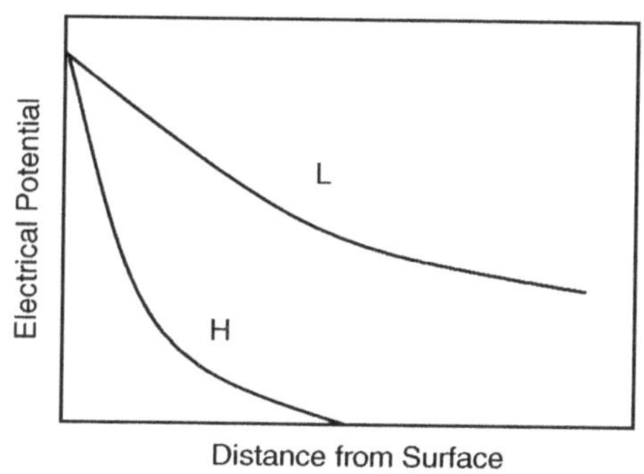

Figure 2.9. A schematic representation for the effect of particle concentration on the Gibbs free energy of the electrostatic interaction. The symbols H and L represent the high and low levels of particle concentration, respectively.

surface charge density, limited flocculation of latex particles often occurs and this phenomenon greatly reduces the number of particles per unit volume of water (or increases the particle size) with the progress of polymerization. Accompanied by the limited particle flocculation process, the latex particle surface charge density increases significantly; thus, satisfactory colloidal stability of the polymerization system can be achieved. The particle sizes of latex products prepared by the surfactant-free emulsion polymerization technique are quite large. A small amount of functional comonomers (e.g., <5 wt% acrylic acid or methacrylic acid based on total monomer weight) can be incorporated into the surfactant-free emulsion polymerization system in order to reduce the resultant latex particle size. At $pH > 5$ (e.g., adjusted by a buffer reagent sodium bicarbonate or ammonium hydroxide), a significant fraction of ionized monomeric units of acrylic acid or methacrylic acid are distributed near the surface layer of latex particles. This then leads to the enhanced latex particle surface charge density and reduced particle size during polymerization.

2.3.3 Steric Interactions

In addition to ionic surfactants, nonionic surfactant molecules can also adsorb onto the particle surfaces to impart satisfactory stability to colloidal dispersions [20, 21]. Some very old examples include India ink and carbon black particles dispersed in the continuous aqueous phase containing a natural gum. This kind of colloidal stabilization mechanism (termed steric stabilization) was first illustrated experimentally by M. Faraday [31, 32]. Some representative polymeric materials (protective colloids) that are effective in preparing sterically stabilized aqueous colloidal dispersions are summarized in Table 2.7 [21]. A portion of an effective protective colloid must be hydrophobic enough to show a strong tendency to adsorb onto the hydrophobic particle surface. Furthermore, the adsorbed macromolecules must form a relatively thick hydrophilic layer surrounding the particle, which serves as a steric barrier to prevent the colloidal particles from flocculation.

As discussed in Chapter 1 (Section 1.3.3), the steric stabilization effect is the resultant of the reduction in entropy when two hairy particles approach each other. The primary parameters that govern the steric interactions between an isolated pair of particles include (a) the surface concentration of polymer chains adsorbed onto the particle, (b) the thickness of the adsorbed layer of polymer chains, and (c) the concentration profile of the outer part of the adsorbed layer of polymer chains.

The first attempt to develop a theoretical model for calculating the Gibbs free energy of the steric interaction (ΔG_s) between an isolated pair of flat plates was reported by Mackor [33]. This model considers changes in entropy when two well-separated flat plates with rod-like polymer chains adsorbed on

Table 2.7. Some Representative Polymeric Moieties Used to Prepare Stable Colloidal Dispersions [21]

Aqueous Dispersions	
Disperse Phase	Protective Colloid
Polystyrene	Polyethylene glycol
Polyvinyl acetate	Polyvinyl alcohol
Polymethyl methacrylate	Polyacrylic acid
Polyacrylonitrile	Polymethacrylic acid
Polydimethylsiloxane	Polyacrylamide
Polyvinyl chloride	Polyvinyl pyrrolidone
Polyethylene	Polyethylene imine
Polypropylene	Polyvinyl methyl ether
Polydodecyl methacrylate	Poly4-vinylpyridine

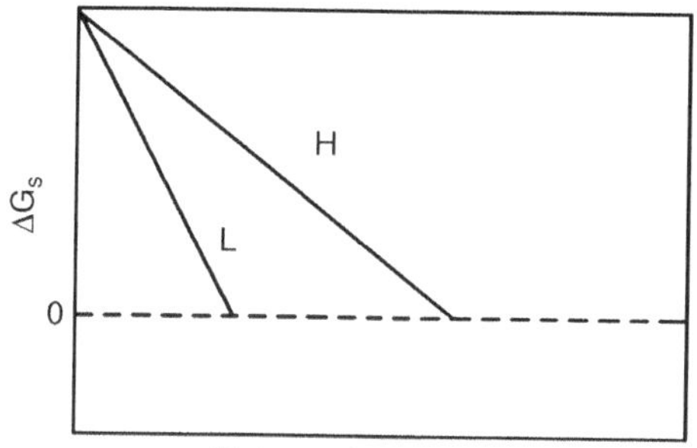

Figure 2.10. A schematic representation of the effect of the adsorbed polymer chain length on the Gibbs free energy of the steric interaction. The symbols H and L represent the high and low levels of the adsorbed polymer chain length, respectively.

them are brought together to the extent that the number of possible configurations of the adsorbed polymer chain is reduced. The expression for ΔG_s obtained from this simple approach that is illustrative of the basic principle of the steric stabilization mechanism is

$$\Delta G_s = NkT\theta[1-(H/l)] \tag{2.27}$$

where N is the number of the adsorbed polymer chains per unit surface area, θ is the surface coverage of polymer chains when H approaches infinity, H is the distance of separation between the plates, and l is the length of the adsorbed polymer chain. Equation (2.27) predicts that ΔG_s increases with increasing N, H, or l. Figure 2.10 shows a schematic representation of the effect of the adsorbed polymer chain length on ΔG_s.

Based on the excess osmotic pressure established by the difference in the concentration of the adsorbed polymer chains between the overlap region and the unaffected region of an isolated pair of particles in combination with the Flory–Huggins polymer solution theory [34–38], for constant polymer chain segment density, a more realistic model developed for calculating ΔG_s is shown below [6, 21, 39]:

$$\Delta G_s = 4\pi kTa\omega^2 N_A(v_2^2/V_1)(0.5-\chi)\{1-[H_0/(2\delta)]\}^2 \tag{2.28}$$

where ω is the mass of the adsorbed polymer chains per unit surface area, V_1 is the molar volume of water, and v_2 is the partial specific volume of the adsorbed polymer, N_A is Avogadro's number, χ is the interaction parameter between polymer and water, H_0 is the minimum distance between the surfaces of the interactive particles, and δ is the thickness of the adsorbed layer of polymer chains. The excess osmotic pressure is the driving force for the generation of the repulsive steric interaction between the close approaching particles. Furthermore, Eq. (2.28) predicts that the Gibbs free energy of the repulsive steric interaction is linearly proportional to the particle radius (a) and the thermodynamic factor ($0.5 - \chi$), and it is also proportional to the square of the number density of the adsorbed polymer chains (ω) or the geometric factor ($\{1 - [H_0/(2\delta)]\}$). These dependences reflect some general features of the sterically stabilized colloidal dispersions as follows:

(a) Steric stabilization is very sensitive to changes in temperature. In aqueous colloidal dispersions, χ approaches 0.5 as the temperature approaches the lower critical solution temperature (LCST) of the adsorbed polymer. This will then lead to a value of ΔG_s close to zero and then the loss of colloidal stability. For example, the hydrogen bonding strength between polyethylene glycol (PEG), commonly used as the hydrophilic part of nonionic surfactants, and water decreases with increasing temperature. As a result, the adsorbed PEG chains surrounding the particle shrink and then become less effective in stabilizing the colloidal dispersion.

(b) The repulsive steric interaction increases rapidly with increasing concentration of the adsorbed polymer chains within the particle surface layer.

(c) The repulsive steric interaction increases significantly with decreasing distance of separation (i.e., increasing overlap volume) between the particles.

The molecular weight of the adsorbed polymer chains around the particle may also have an influence on the colloidal stability. Intuitively, the higher the polymer molecular weight, the thicker particle surface layer of the adsorbed polymer chains and, therefore, the better the colloidal stability. However, if a very-high-molecular-weight polymer with multiple potential points for adsorption is used to stabilize the colloidal dispersion, the polymer chain may become adsorbed onto several particles and thereby lead to bridging flocculation [40]. This is especially true when the population of colloidal particles is relatively small (Figure 2.11).

Another unique phenomenon involving colloidal dispersions stabilized by low molecular weight, weakly adsorbed polymer chains is the depletion flocculation mechanism [41], as shown in Figure 2.12. When an isolated pair of the particles approach each other, the weakly adsorbed polymer chains are squeezed out of the overlap volume due to the greatly reduced space available for these polymer chains. This then results in the imbalance of the local osmotic pressure; that is, the concentration of the adsorbed polymer is lower than that in the continuous bulk phase. Thus, water molecules are forced to diffuse out of the overlap region to counterbalance the osmotic pressure effect. The net effect is that the particles are pulled together and flocculation takes place.

The Gibbs free energy of interactions between an isolated pair of particles (ΔG) for the colloidal system stabilized by both the electrostatic and steric interactions then can be expressed as

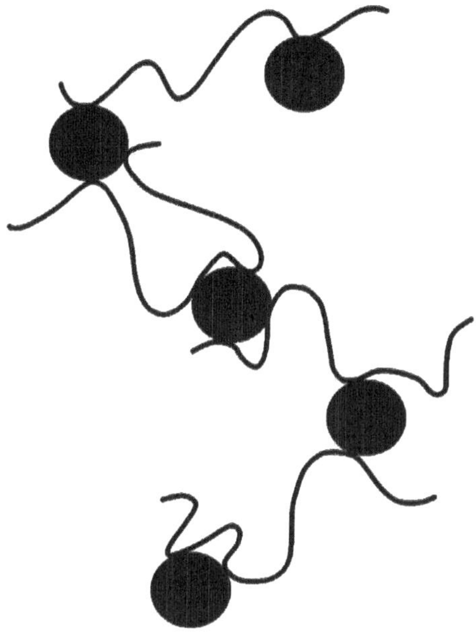

Figure 2.11. A schematic model for the bridging flocculation mechanism.

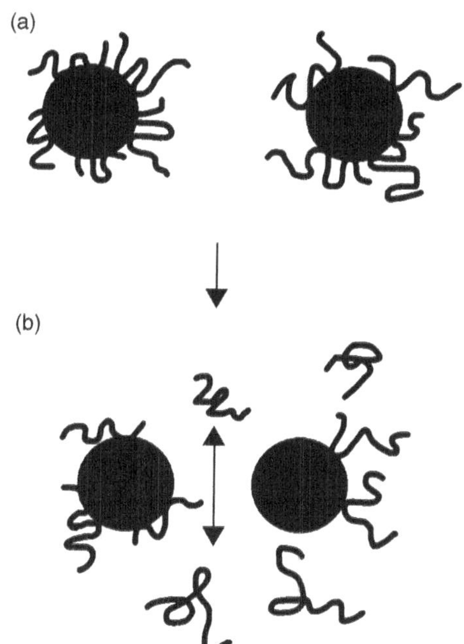

Figure 2.12. A schematic model for the depletion flocculation process. **(a)** Initially an isolated pair of the particles are stabilized by the low-molecular-weight, weakly adsorbed polymer chains. **(b)** Depletion flocculation occurs when these two particles approach to each other and the overlap volume is exceedingly reduced.

Table 2.8. Characteristics of the Electrostatically and Sterically Stabilized Colloidal Dispersions [39]

Electrostatic Stabilization	Steric Stabilization
Sensitive to electrolyte	Insensitive to electrolyte
Mainly effective in aqueous dispersion	Equally effective in aqueous and nonaqueous dispersions
More effective at low solids content	Equally effective at high and low solids contents
Often irreversible coagulation	Generally reversible flocculation
Poor freeze–thaw stability	Good freeze–thaw stability

$$\Delta G = \Delta G_v + \Delta G_e + \Delta G_s \tag{2.29}$$

Table 2.8 summarizes the quite different characteristics of the electrostatically and sterically stabilized colloidal dispersions [39]. This may explain why the synergistic effect of the mixed ionic and nonionic surfactants often results in a better colloidal stability.

2.3.4 Kinetics of Flocculation

Colloidal particles tend to flocculate with one another when the attractive van der Waals interaction term predominates in Eq. (2.29). It is of great interest to gain a fundamental understanding of the kinetics of flocculation whether the instability of colloidal dispersions is desirable or not. The pioneering work of von Smoluchowski [42–44] dealing with the diffusion-controlled kinetics of flocculation (i.e., formation of a doublet immediately after collision of the particles) starts with the following expression for the flux (J) of the incoming particle toward the central particle acting as a sink.

$$J = N_p(2D/a)4\pi(2a)^2 \tag{2.30}$$

where N_p is the number of primary particles per unit volume of water, D is the diffusion coefficient, and $4\pi(2a)^2$ represents the surface area of the collision sphere. According to the Stokes–Einstein equation, the value of D can be calculated as follows:

$$D = kT/6\pi\eta_0 a \tag{2.31}$$

where η_0 is the viscosity of the continuous aqueous phase. The detailed model development will not be discussed here, and only the resultant second-order governing equation with the initial condition (I.C.) is presented as follows:

$$\begin{aligned} dN_p/dt &= -(8kT/3\eta_0)N_p^2 \\ \text{I.C.:} \quad N_p &= N_{p0} \quad \text{when } t = 0 \end{aligned} \tag{2.32}$$

where N_{p0} is the number of primary particles per unit volume of water initially present in the colloidal system. The analytical solution obtained from such a simple approach is given below.

$$\begin{aligned} 1/N_p &= 1/N_{p0} + (8kT/3\eta_0)t \\ &= 1/N_{p0} + k_f t \end{aligned} \quad (2.33)$$

$$t_{1/2} = 3\eta_0/(4kTN_{p0}) \quad (2.34)$$

where k_f defined as $(8kT/3\eta_0)$ is the fast particle flocculation rate constant and $t_{1/2}$ is the half-life of the diffusion-controlled flocculation process. For a typical dilute aqueous colloidal dispersion at 25 °C ($N_{p0} \sim 10^{15}\,\mathrm{dm}^{-3}$), the value of $t_{1/2}$ was estimated to be 10^{-1} s, which is shorter than the time (seconds to minutes) normally observed in practice for the dispersions in the absence of any stabilization mechanism [5]. This can be attributed to the fact that water molecules between the approaching particles must move out of the way before particle flocculation can occur. Furthermore, colloidal dispersions (e.g., latex products) should have a shelf life of at least several months in most industrial applications. Apparently, this cannot be achieved without resort to some kind of stabilization mechanisms such as electrostatic or steric interactions.

Equation (2.33) tends to overestimate the particle flocculation rate for the electrostatically or sterically stabilized colloidal dispersions due to the potential energy barrier established between the approaching particles against particle flocculation. Fuchs [45] was the first to introduce the stability ratio (W) to take into account the fact that not all the collision events lead to successful coagulation of colloidal particles.

$$W = 2a\int_{2a}^{\infty} \{\exp[\Delta G/(kT)]/r^2\}dr \quad (2.35)$$

The reciprocal of the stability ratio ($1/W$) simply represents the fraction of collisions that effectively results in the formation of doublets. The larger the Gibbs free energy of interactions (ΔG), the larger the stability ratio (W). As a result, a more stable colloidal system is achieved. Under these circumstances, the particle flocculation rate can be written as

$$\begin{aligned} dN_p/dt &= -(k_f/W)N_p^2 \\ &= -k_s N_p^2 \\ \text{I.C.:} \quad N_p &= N_{p0} \quad \text{when } t = 0 \end{aligned} \quad (2.36)$$

where k_s ($=k_f/W$) is the slow particle flocculation rate constant. Equation (2.36) is readily reduced to the von Smoluchowski model [Eq. (2.32)–(2.34)] when W approaches one.

Reerink and Overbeek [46] showed that the maximum Gibbs free energy of interactions between an isolated pair of particles (ΔG_{max}) predominates in the slow particle flocculation process and the stability ratio (W) can be estimated by the following equation:

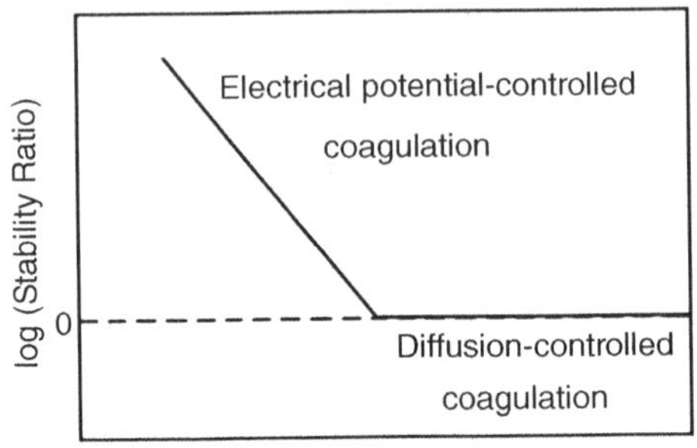

Figure 2.13. A schematic plot for the stability ratio as a function of the concentration of counterion.

$$W \sim 1/(2\kappa a)\exp(\Delta G_{\max}/kT) \tag{2.37}$$

Equation (2.37) predicts that increasing κ (or concentration of counterion) results in a significant reduction in both the preexponential term ($1/(2\kappa a)$) and $\Delta G_{\max}$. It was also proposed that the influence of the electrolyte concentration (C) on the stability ratio can be described by the following equation:

$$\log W = -k_1 \log C + k_2 \tag{2.38}$$

where k_1 and k_2 are the kinetic parameters for the slow particle flocculation process. This forms the basis of coagulation kinetics experiments with the light-scattering technique used to evaluate the sensitivity of colloidal dispersions to added electrolyte and for determination of their CCC values (see the $\log W$ versus $\log C$ profile in Figure 2.13). The logarithm of W decreases linearly with increasing logarithm of electrolyte concentration in the slow particle flocculation region. The magnitude of k_1 obtained from the least squares best-fitted slope of the $\log W$ versus $\log C$ straight line reflects the stability of colloidal dispersions. The larger the absolute value of k_1, the stronger the tendency for the particles to undergo flocculation. The point at which $W = 1$ (i.e., no potential energy barrier between the particles against coagulation) is then identified as the critical coagulation concentration (CCC) of electrolyte. Beyond the critical coagulation concentration, the particle flocculation process becomes diffusion-controlled.

For those who are interested in the details of the particle flocculation kinetics, refer to references 2, 18, 19, and 39.

REFERENCES

1. S. Ross and I. D. Morrison, *Colloidal Systems and Interfaces*, John Wiley & Sons, New York, 1988.

2. W. B. Russel, D. A. Saville, and W. Schowalter, *Colloidal Dispersions*, Cambridge University Press, Cambridge, 1989.
3. A. W. Adamson, *Physical Chemistry of Surfaces*, 5th ed., John Wiley & Sons, New York, 1990.
4. R. J. Hunter, *Introduction to Modern Colloid Science*, Oxford University Press, New York, 1993.
5. D. Myers, *Surfaces, Interfaces, and Colloids: Principles and Applications*, 2nd ed., John Wiley & Sons, New York, 1999.
6. J. W. Goodwin, *Colloids and Interfaces with Surfactants and Polymers—An Introduction*, John Wiley & Sons, West Sussex, 2004.
7. W. D. Harkins, *J. Chem. Phys.* **13**, 381 (1945).
8. W. D. Harkins, *J. Chem. Phys.* **14**, 47 (1946).
9. W. D. Harkins, *J. Am. Chem. Soc.* **69**, 1428 (1947).
10. W. V. Smith, *J. Am. Chem. Soc.* **70**, 3695 (1948).
11. W. V. Smith and R. H. Ewart, *J. Chem. Phys.* **16**, 592 (1948).
12. W. V. Smith, *J. Am. Chem. Soc.* **71**, 4077 (1949).
13. W. C. Griffin, *J. Soc. Cosmet. Chem.* **5**, 249 (1954).
14. J. T. Davies and E. K. Rideal, *Interfacial Phenomena*, Academic Press, New York, 1963.
15. A. F. M. Barton, *CRC Handbook of Solubility Parameters and Other Cohesion Parameters*, CRC Press, Boca Raton, FL, 1983, pp. 142–149, 445.
16. P. A. Small, *J. Appl. Chem.* **3**, 71 (1953).
17. B. V. Derjaguin and L. Landau, *Acta Physicochim.* **14**, 633 (1941).
18. E. J. W. Verwey and J. Th. G. Overbeek, *Theory of the Stability of Lyophobic Colloids*, Elsevier, Amsterdam, 1948.
19. P. C. Hiemenez, *Principles of Colloids and Surface Chemistry*, 2nd ed., Marcel Dekker, New York, 1986.
20. T. Sato and R. Ruch, *Stabilization of Colloidal Dispersions by Polymer Adsorption*, Marcel Dekker, New York, 1980.
21. D. H. Napper, *Polymeric Stabilization of Colloidal Dispersions*, Academic Press, London, 1983.
22. B. W. Ninham and V. A. Parsegian, *Biophys. J.* **10**, 646 (1970).
23. V. A. Parsegian and G. H. Weiss, *J. Colloid Interface Sci.* **81**, 285 (1981).
24. D. B. Hough and L. R. White, *Adv. Colloid Interface Sci.* **14**, 3 (1980).
25. J. N. Israelachvili, *Intermolecular and Surface Forces*, Academic Press, New York, 1985.
26. M. D. Croucher and M. L. Hair, *J. Phys. Chem.* **81**, 1631 (1977).
27. D. Tabor and R. H. S. Winterton, *Proc. Roy. Soc.* **A312**, 435 (1969).
28. A. B. Glendinning and W. B. Russel, *J. Colloid Interface Sci.* **93**, 95 (1983).
29. R. Hogg, T. W. Healy, and D. W. Fursteneau, *Trans. Faraday Soc.* **62**, 1638 (1966).
30. J. Th, G. Overbeek, *Pure Appl. Chem.* **52**, 1151 (1980).
31. M. Faraday, *Phil. Trans.* **147**, 145 (1857).
32. E. Hatschek, *The Foundations of Colloid Science*, Ernest Benn, London, 1925, pp. 89–91.

33. E. L. Mackor, *J. Colloid Interface Sci.* **6**, 492 (1951).
34. P. J. Flory, *J. Chem. Phys.* **9**, 660 (1941).
35. P. J. Flory, *J. Chem. Phys.* **10**, 51 (1942).
36. M. L. Huggins, *J. Chem. Phys.* **9**, 440 (1941).
37. M. L. Huggins, *J. Phys. Chem.* **46**, 151 (1942).
38. M. L. Huggins, *J. Am. Chem. Soc.* **64**, 1712 (1942).
39. R. J. Hunter, *Foundations of Colloid Science*, Vol. I, Oxford University Press, New York, 1989.
40. R. A. Ruehrwein and D. W. Ward, *Soil Sci.* **73**, 485 (1952).
41. S. Asakura and F. Oosawa, *J. Chem. Phys.* **22**, 1255 (1954).
42. M. von Smoluchowski, *Physik. Z.* **17**, 557 (1916).
43. M. von Smoluchowski, *Physik. Z.* **17**, 585 (1916).
44. M. von Smoluchowski, *Z. Phys. Chem.* **92**, 129 (1917).
45. N. Fuchs, *Z. Physik.* **89**, 736 (1934).
46. H. Reerink and J. Th. G. Overbeek, *Discuss. Faraday Soc.* **18**, 74 (1954).

3

PARTICLE NUCLEATION MECHANISMS

In conventional emulsion polymerization, free radical chain polymerization primarily takes place in discrete monomer-swollen polymer particles (i.e., reaction loci, about 10^1–10^2 nm in diameter) dispersed in a continuous aqueous phase. The heterogeneous reaction system initially comprises monomer droplets and monomer-swollen micelles and a continuous aqueous phase saturated with monomer and surfactant. Immediately after the addition of a water-soluble initiator [e.g., sodium persulfate (NaPS)] that is thermally decomposed to generate a pair of free radicals in the continuous aqueous phase, particle embryos start to form in the polymerization system. The particle nucleation period is generally quite short, but the resultant number and size of these polymer particles per unit volume of water govern the polymerization kinetics and have a significant influence on the performance properties (e.g., rheology, film formation, gloss, etc.) of latex products. The rate of polymerization is linearly proportional to the number density of particles and the average number of free radicals per particle. The latter is closely related to the size of the growing particles and transport of free radicals in the heterogeneous polymerization system.

For a typical semibatch emulsion polymerization system, the initial reactor charge comprises water, surfactants, and sometimes a small proportion of monomers. When the reaction temperature (e.g., 80 °C) is reached, a persulfate initiator solution is added to the initial reactor charge to generate free radicals. This is then followed by the continuous addition of monomers (or monomer emulsion) over a period of time (normally a few hours). The appearance of the reaction mixture is transformed from transparent into translucent, opaque

Principles and Applications of Emulsion Polymerization, by Chorng-Shyan Chern

and then milky white with the progress of polymerization, which reflects the nucleation and growth of latex particles. How do these particle nuclei form and grow during the emulsion polymerization? Where is the locus of particle nucleation, the continuous aqueous phase, monomer droplets (>10° μm in diameter) or monomer-swollen micelles (~10° nm in diameter)? What are the primary reaction parameters that control the particle nucleation and growth processes? What is the best strategy to manipulate particle nucleation and growth processes in order to achieve the target particle size and particle size distribution with minimum levels of surfactants and filterable solids in the design of a latex product? These critical issues have been the focus of many studies in the last half a century. Several books and review articles dealing with emulsion polymerization in the last decade are listed in references 1–8.

For conciseness, this chapter primarily deals with three well-established particle nucleation mechanisms (i.e., micellar nucleation, homogeneous nucleation and coagulative nucleation). This is followed by the discussion of emulsion polymerization kinetics in Chapter 4.

3.1 MICELLAR NUCLEATION

3.1.1 Harkins–Smith–Ewart Theory

The pioneering work of Harkins [9–11], Smith and Ewart [12–14], and Gardon [15, 16] disclosed the general features of emulsion polymerization. A typical recipe comprises relatively hydrophobic monomers (e.g., styrene and butadiene), water, surfactant, and a water-soluble persulfate initiator. The reaction system is characterized by emulsified monomer droplets (>10° μm in diameter, 10^{12}–10^{14} dm^{-3} in number) dispersed in a continuous aqueous phase with the aid of the oil-in-water type of surfactant at the very beginning of polymerization (see Figure 1.4a). Monomer-swollen micelles (about 5–10 nm in diameter, 10^{19}–10^{21} dm^{-3} in number) may also exist in the reaction system if the concentration of surfactant in the continuous aqueous phase is above its critical micelle concentration (CMC). Only a small fraction of monomer molecules are present in the micelles (if present) or dissolved in the continuous aqueous phase. Most of the monomer molecules dwell in the giant monomer reservoirs (i.e., monomer droplets). The free radical polymerization is initiated by the addition of initiator. It was proposed that latex particles (about 10^{-2}–10^{1} μm in diameter, 10^{16}–10^{18} dm^{-3} in number) are generated via the capture of free radicals generated in the continuous aqueous phase by monomer-swollen micelles, which exhibit an extremely large oil–water interfacial area. In general, monomer droplets are not effective in competing with micelles in capturing free radicals due to their relatively small droplet surface area.

Waterborne free radicals first polymerize with monomer molecules dissolved in the continuous aqueous phase. This would result in an increase in the hydrophobicity of oligomeric radicals. When the critical chain length is achieved, these oligomeric radicals become so hydrophobic that they show a strong tendency to enter the monomer-swollen micelles and then continue to react with those monomer molecules therein. As a consequence, the monomer-swollen micelles attacked by oligomeric radicals are successfully transformed into particle nuclei. These embryo particles continue to grow by acquiring monomer from monomer droplets and monomer-swollen micelles. In order to maintain adequate colloidal stability of the growing particle nuclei, micelles that do not contribute to particle nucleation disband to supply the increasing demand for surfactant. In addition, the surfactant molecules adsorbed on monomer droplet surfaces may also desorb from the droplet surfaces, diffuse across the continuous aqueous phase, and then adsorb on the expanding particle surfaces. The particle nucleation stage (Interval I) ends immediately after the exhaustion of monomer-swollen micelles. About one of every 10^2–10^3 monomer-swollen micelles can be successfully converted into latex particles.

Based on the above reaction scheme and the assumptions that (a) a monomer-swollen micelle can be successfully converted into a particle nucleus via the capture of a free radical in the continuous aqueous phase, (b) the volumetric growth rate for particle nuclei ($\mu = dv_p/dt$, where v_p is the volume of a particle nucleus) is constant, (c) desorption of free radicals out of the particle does not occur, and (d) the amount of surfactant molecules dissolved in the continuous aqueous phase and adsorbed on the monomer droplet surfaces is insignificant, the rate of formation of particle nuclei is then equal to the rate of generation of free radicals in the continuous aqueous phase.

$$dN_p/dt = 2fk_d[\mathrm{I}] \tag{3.1}$$

$$d[\mathrm{I}]/dt = -k_d[\mathrm{I}] \tag{3.2}$$

where N_p is the number of latex particles per unit volume of water, t is the reaction time, f is the initiator efficiency factor ($0 \leqq f \leqq 1$), k_d is the initiator decomposition rate constant, and $[\mathrm{I}]_0$ and $[\mathrm{I}]$ ($=[\mathrm{I}]_0 \exp(-k_d t)$) are the initiator concentrations at time equal to zero and t, respectively. The value of $[\mathrm{I}]$ is approximately equal to that of $[\mathrm{I}]_0$ during the early stage of polymerization. For example, the half-life of a persulfate initiator at 80°C is about 2 hours. Equation (3.1) predicts a relatively constant rate of particle nucleation because this interval is generally quite short.

Particle nucleation stops immediately after the depletion of monomer-swollen micelles, and the corresponding time is represented by the symbol t_I. At t_I, the volume of a particle nucleus generated at time τ ($v_{p,\mathrm{I}}$) is $\mu(t_\mathrm{I} - \tau)$ and the corresponding particle surface area ($a_{p,\mathrm{I}}$) of this nucleus is then $[(4\pi)^{1/2}3\mu]^{2/3}(t_\mathrm{I} - \tau)^{2/3}$. Thus, the total particle surface area at t_I ($A_{p,\mathrm{I}}$) is $\int[(4\pi)^{1/2}3\mu]^{2/3}(t_\mathrm{I}-\tau)^{2/3}(2fk_d[\mathrm{I}])\,d\tau$ (integrated from 0 to t_I) $= \frac{3}{5}[(4\pi)^{1/2}3\mu]^{2/3}(2fk_d[\mathrm{I}])t_\mathrm{I}^{5/3}$ which is equal to $a_s S_0$. The parameter a_s is the particle surface area occupied by unit weight of the adsorbed surfactant, and S_0 is the weight of surfactant initially present in the polymerization system. In this manner, the time required to complete particle nucleation t_I is equal to $0.53\mu^{-2/5}[a_sS_0/(2fk_d[I])]^{3/5}$. As expected, the length of the particle nucleation period decreases with decreasing the surfactant concentration, increasing the initiator concentration or increasing the particle growth rate. The total number of monomer-swollen polymer particles generated at the end of Interval I ($N_{p,I}$) can be calculated by the following equation because all the surfactant molecules present in the polymerization system are used to stabilize the growing particle nuclei.

$$\begin{aligned} N_{p,I} &= (2fk_d[I])t_I \\ &= 0.53[(2fk_d[I]/\mu)]^{2/5}(a_sS_0)^{3/5} \end{aligned} \tag{3.3}$$

Thus, Eq. (3.3) predicts that the total number of particle nuclei generated in Interval I is proportional to the initiator concentration and the surfactant concentration to the 0.4 and 0.6 powers, respectively, as shown schematically in Figure 3.1. This relationship was confirmed by the experimental data obtained from emulsion polymerization of relatively hydrophobic monomers such as styrene. This suggests that the concentration of surfactant plays an important role in the particle nucleation process.

Considering a more realistic scenario that the existing particle nuclei may also compete with monomer-swollen micelles in capturing free radicals, the rate of particle nucleation then becomes

$$dN_p/dt = 2fk_d[I][1-(A_p/(a_sS_0))] \tag{3.4}$$

where A_p is the total particle surface area during the particle nucleation stage (Interval I). Similarly, an expression for the calculation of $N_{p,I}$ can be derived as follows:

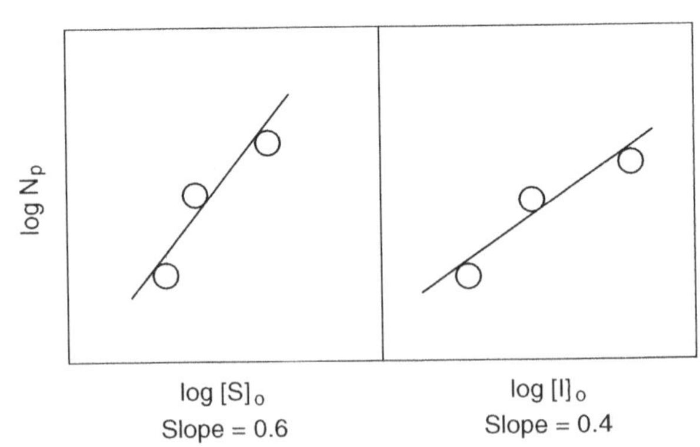

Figure 3.1. A schematic representation of the emulsion polymerization system that follows the Smith–Ewart theory. The symbols $[S]_0$ and $[I]_0$ represent the concentrations of surfactant and initiator, respectively, initially present in the reaction system.

$$N_{p,1} = 0.37[(2fk_d[\mathrm{I}])/\mu]^{2/5}(a_s S_0)^{3/5} \tag{3.5}$$

In comparison with Eq. (3.3), a smaller population of latex particles ($N_{p,1}$) is obtained from Eq. (3.5) due to the competitive absorption of free radicals.

At the very beginning of polymerization, the total particle surface area is much smaller than the total micelle surface area because of the very low number density of particle nuclei produced. Thus, the effect of the existing particle nuclei on the fate of free radicals is negligible and Eq. (3.4) can be reduced to Eq. (3.1). On the other hand, the total particle surface area becomes so large that free radicals may also have the chance to enter the existing particle nuclei during the latter stage of Interval I. As a result, the rate of particle nucleation is greatly reduced. Equations (3.3) and (3.5) represent the upper limit and lower limit, respectively, of the total number of latex particles per unit volume of water that can be generated in a particular emulsion polymerization system.

It should be noted that the derivation of Eq. (3.5) is based on the assumption that free radicals are captured by monomer-swollen micelles or particle nuclei at a rate that is proportional to their surface area (termed the collision theory [15]). In this case, there is no free radical concentration gradient surrounding the colloidal particle. It would be more appropriate to use the diffusion theory [17, 18] to calculate the rate of entry of free radicals into micelles or particle nuclei if this concentration gradient does exist. The diffusion theory proposes that the rate of entry of free radicals into a colloidal particle is equal to $2\pi d_p D_w[\mathrm{R}^*]_w$, where d_p is the diameter of the particle, D_w is the diffusion coefficient of free radicals in water, and $[\mathrm{R}^*]_w$ is the bulk concentration of free radicals in water. It is generally accepted that the diffusion theory is more realistic to describe the absorption of free radicals by micelles or particle nuclei [19]. The detailed reaction mechanisms involved in the entry of free radicals into monomer-swollen micelles or particle nuclei will be discussed in Chapter 4.

3.1.2 Competitive Absorption of Free Radicals by Micelles and Particle Nuclei

The concise Harkins–Smith–Ewart theory [9–16] delicately describes the key characteristics of emulsion polymerization. However, the difference in colloidal properties (e.g., composition, size, surface charge density, and particle surface area occupied by the adsorbed surfactant) between the monomer-swollen micelles and particle nuclei was not taken into account in the derivation of Eq. (3.4). The probability for micelles or particle nuclei to capture oligomeric radicals in the continuous aqueous phase is simply assumed to be proportional to their total oil–water interfacial area.

Nomura and co-workers [20] studied the competitive absorption of free radicals by micelles and particle nuclei. In addition to the basic assumptions made in the Harkins–Smith–Ewart theory, the bimolecular termination reaction in the continuous aqueous phase was assumed to be insignificant. A mechanistic model comprising a series of elementary reactions was then developed to simulate the particle nucleation process.

Absorption of Radicals by Micelles

$$m + R^*_{(aq)} \rightarrow N_1, \qquad dN_1/dt = k_{e,m} m [\mathrm{R}^*]_w \tag{3.6}$$

Absorption of Radicals by Active Particles

$$N_1 + \mathrm{R}^*_{(aq)} \rightarrow N_0, \qquad dN_0/dt = k_{e,p} N_1 [R^*]_w \tag{3.7}$$

Absorption of Radicals by Inactive Particles

$$N_0 + \mathrm{R}^*_{(aq)} \rightarrow N_1, \qquad dN_1/dt = k_{e,p} N_0 [R^*]_w \tag{3.8}$$

where m, N_1, and N_0 represent the number of monomer-swollen micelles, particle nuclei containing one free radical (active), and particle nuclei containing no radicals (inactive) per unit volume of water, respectively. $\mathrm{R}^*_{(aq)}$ represents the free radicals in the continuous aqueous phase and $[\mathrm{R}^*]_w$ is the concentration of free radicals in the continuous aqueous phase. The parameters $k_{e,m}$ and $k_{e,p}$ are the rate constants for the absorption of free radicals by micelles and particle nuclei, respectively. Figure 3.2 shows a schematic representation of the competitive absorption of waterborne free radicals by monomer-swollen micelles and particle nuclei.

Based on this particle nucleation mechanism, the mole balance of free radicals in the continuous aqueous phase and the rate of particle nucleation can be described by the following equations.

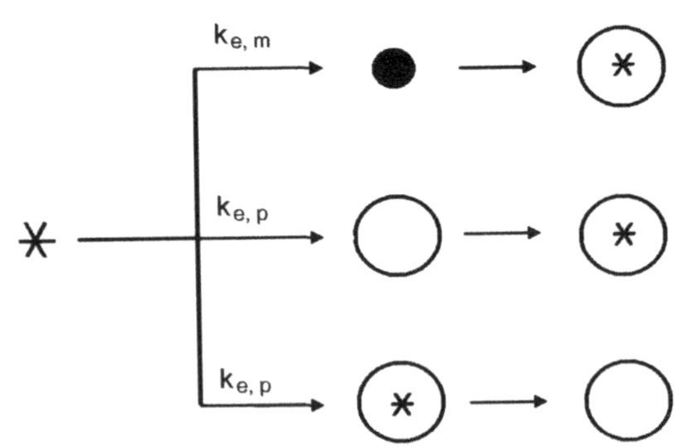

Figure 3.2. A schematic model for the competitive absorption of free radicals by monomer-swollen micelles and particle nuclei. • (~10⁰ nm in diameter), ○ (~10¹ nm in diameter), and ⊛ (~10¹ nm in diameter) represent the monomer-swollen micelles, inactive polymer particle nuclei, and active polymer particle nuclei, respectively. The symbol * represents the free radicals in the continuous aqueous phase.

$$
\begin{aligned}
d[R^*]_w/dt &= 2fk_d[\mathrm{I}] - k_{e,m}m[R^*]_w - k_{e,p}N_1[R^*]_w - k_{e,p}N_0[R^*]_w \\
&= 2fk_d[\mathrm{I}] - k_{e,m}m[R^*]_w - k_{e,p}(N_1 + N_0)[R^*]_w \\
&= 2fk_d[\mathrm{I}] - k_{e,m}m[R^*]_w - k_{e,p}N_p[R^*]_w
\end{aligned} \tag{3.9}
$$

$$
dN_p/dt = k_{e,m}m[R^*]_w \tag{3.10}
$$

Applying the pseudo-steady-state assumption to Eq. (3.9) because of the extremely reactive free radicals and substituting the resultant expression for $[R^*]_w$ into Eq. (3.10), the rate of particle nucleation then becomes

$$
dN_p/dt = 2fk_d[\mathrm{I}]/\{1+[k_{e,p}N_p/(k_{e,m}m)]\} \tag{3.11}
$$

The ratio $k_{e,p}N_p/(k_{e,m}m)$ signifies the competitive absorption of free radicals by micelles and particle nuclei. Only those kinetic events involving the entry of free radicals into micelles contribute to the formation of particle nuclei. Eq. (3.11) is reduced to Equation (3.3) when the ratio $k_{e,p}N_p/(k_{e,m}m)$ is much smaller than one. On the other hand, the rate of particle nucleation can be written as

$$
dN_p/dt = 2fk_d[\mathrm{I}]k_{e,m}m/(k_{e,p}N_p) \tag{3.12}
$$

when the ratio $k_{e,p}N_p/(k_{e,m}m)$ is much greater than one. Based on a set of simultaneous differential equations describing the rate of changes in N_p, N_1, and monomer conversion and a mass balance equation for m, the total number of particle nuclei generated in Interval I is proportional to the initiator and surfactant concentrations to the 0.3 and 0.7 powers, respectively [20].

Taking into account the effect of desorption of free radicals out of the particle nuclei, the following equation was developed to predict the rate of particle nucleation in the emulsion polymerization of relatively hydrophilic monomers.

$$
dN_p/dt = (2fk_d[\mathrm{I}] + k_{\mathrm{des}}\mathbf{n}N_p)/\{1+[k_{e,p}N_p/(k_{e,m}m)]\} \tag{3.13}
$$

where k_{des} is the rate constant for desorption of free radicals out of the particle nuclei and **n** is the average number of free radicals per particle. It was also assumed that only the relatively hydrophilic monomeric radical originating from chain transfer of a propagating radical to monomer is capable of desorbing out of the particle nucleus in the derivation of Eq. (3.13). The number of monomer-swollen micelles initially present in the polymerization system (m) can be estimated by the relationship $m = S_m/n_{\mathrm{agg}}$, where S_m is the total number of surfactant molecules that contributes to the formation of micelles and n_{agg} is the aggregation number of surfactant molecules per micelle. The only unknown parameters in Eq. (3.13) are $k_{e,p}$ and $k_{e,m}$.

For emulsion polymerization of the relatively hydrophilic vinyl acetate, Eq. (3.13) predicts that the total number of particle nuclei generated in Interval I is proportional to the initiator and surfactant concentrations to the 0.04 and 0.94 powers, respectively [21]. This result strongly suggests that the polarity of monomers has a significant influence on the total number of particle nuclei per unit volume of water produced in the particle nucleation stage. However, in comparison with the theoretical value predicted by the diffusion-controlled entry of free radicals into micelles or particle nuclei ($k_{e,p}/k_{e,m} = d_p/d_{\mathrm{m}} \sim 10/1$ and $n_{\mathrm{agg}} \sim 100$, where d_p and d_m are the diameters of particle nuclei and micelles, respectively) [18], an unusually high value of $(k_{e,p}/k_{e,m})n_{\mathrm{agg}}(1.2 \times 10^7)$ was required to satisfactorily fit the experimental data [18]. The deviation between the calculated and experimental values of $(k_{e,p}/k_{e,m})n_{\mathrm{agg}}$ was attributed to the free radical capture efficiency. This phenomenon was also reported in references 22–24.

Taking into account the effect of desorption of free radicals and the concept of free radical capture efficiency on the formation of particle nuclei, the rate of particle nucleation can be expressed as follows [25]:

$$dN_p/dt = (2fk_d[\mathrm{I}] + k_{\mathrm{des}}\mathbf{n}N_p)\{\eta m d_m^x/[(\eta m d_m^x) + (N_p d_p^x)]\} \tag{3.14}$$

where $\eta(=(k_{e,m}/k_{e,p})\,(d_p/d_{\mathrm{m}})^x)$ is the free radical capture efficiency of a micelle relative to a particle nucleus. The condition of x equal to one is referred to the diffusion-controlled entry of free radicals into micelles or particle nuclei [see Eq. (3.13)]. Assuming that the theory of diffusion-controlled entry of free radicals into micelles or particle nuclei is applicable to the emulsion polymerization of styrene, methyl methacrylate, vinyl acetate or vinyl chloride, the following relationship for the total number of particle nuclei produced in Interval I was obtained.

$$N_p \sim [\mathrm{I}]^{1-z}[S_0]^z \tag{3.15}$$

where 0.6 (a common value for styrene) $< z < 1$ (a common value for vinyl acetate). In general, the effect of desorption of free radicals out of the particle nuclei increases with the increasing rate of chain transfer of a polymeric radical to the monomer or added chain transfer agent in the particle nuclei and the solubility of the resulting free radical in water. The stronger the effect of desorption of free radicals, the larger the value of z.

3.2 HOMOGENEOUS NUCLEATION

3.2.1 Formation of Particle Nuclei in the Continuous Aqueous Phase

Roe [26] pointed out that the derivation of Eqs. (3.3) and (3.5) did not require the assumption of particle nucleation in monomer-swollen micelles. With the assumption that particle nuclei formed outside the micelles and particle nucleation ceased when the particle nuclei grew to such a size that the concentration of surfactant in the continuous aqueous phase was below its CMC, he was able to obtain exactly the same expressions. Another drawback with the micellar nucleation mechanism is that the presence of monomer-swollen micelles is a prerequisite to the nucleation of particle embryos therein. Nevertheless, there is no doubt that particle nuclei can form at a surfactant level lower than its CMC or even in the absence of surfactant (termed the surfactant-free emulsion polymerization). Thus, some mechanisms other than micellar nucleation must be responsible for the particle formation process when the surfactant concentration is below its CMC. For those emulsion polymerization systems stabilized by a level of surfactant lower than its CMC, the resultant latex particle size and particle size distribution are very often quite large and narrow, respectively, and the levels of filterable solids or reactor scraps may become significant, especially for surfactant-free emulsion polymerization systems. However, emulsion polymers containing relatively low levels of surfactant are sometimes desirable because they offer excellent coating properties such as water resistance and adhesion.

Priest [27], Roe [26], and Fitch and Tsai [28–30] proposed the homogeneous nucleation mechanism for the formation of particle nuclei in the continuous aqueous phase, as shown schematically in Figure 3.3. First, waterborne initiator radicals are generated by the thermal decomposition of initiator, and they can grow in size via the propagation reaction with those monomer molecules dissolved in the continuous aqueous phase. The oligomeric radicals then become water-insoluble when the critical chain length is reached. The hydrophobic oligomeric radical may thus coil up and form a particle nucleus in the continuous aqueous phase. This is followed by formation of stable primary particles via the limited flocculation of the relatively unstable particle nuclei to reduce the total oil–water interfacial area and adsorption of surfactant molecules on their particle surfaces to increase the particle surface charge density. The surfactant molecules required to stabilize these primary particles come from those dissolved in the continuous aqueous phase and those adsorbed on the emulsified monomer droplet surfaces.

The above ideas were then incorporated into the following kinetic model developed by Fitch and Tsai [28–30]:

$$dN_p/dt = b\rho_i - R_c - R_f \tag{3.16}$$

where t is the reaction time, ρ_i ($=2fk_d[\mathrm{I}]$) is the rate of generation of free radicals in the continuous aqueous phase, b is an adjustable parameter that takes into account the aggregation of oligomeric radicals, R_c is the rate of capture of free radicals by the particle nuclei, and R_f is the rate of flocculation of the particle nuclei.

The principle behind the homogeneous nucleation mechanism [see Eq. (3.16)] is that the rate of generation of particle embryos is primarily governed

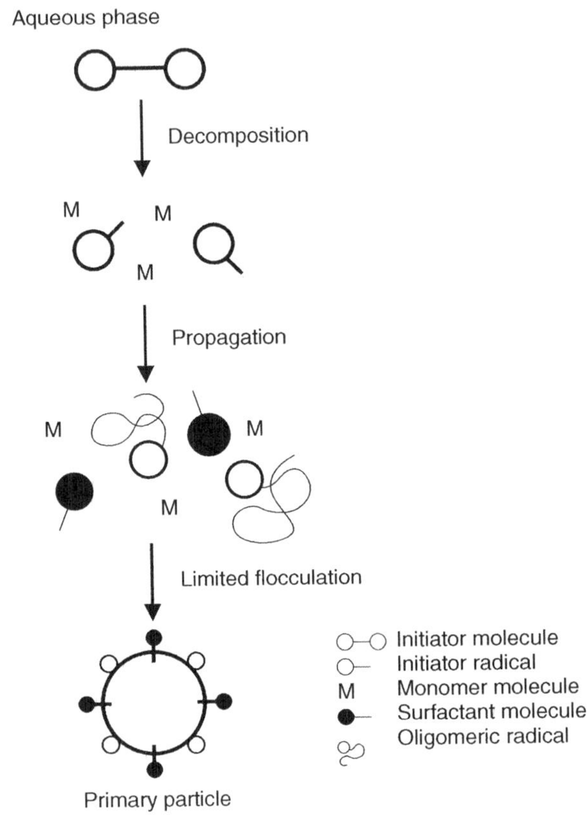

Figure 3.3. A schematic representation of the homogeneous nucleation mechanism.

by the concentration of initiator and reaction temperature. The amounts of surfactant molecules and sulfate end-groups of oligomeric radicals available for stabilizing particle nuclei produced in the continuous aqueous phase control the extent of flocculation of particle nuclei. Furthermore, particle nuclei may also absorb oligomeric radicals from the continuous aqueous phase and, consequently, reduces the rate of generation of particle embryos. The rate of particle nucleation ceases (i.e., $dN_p/dt = 0$) when $b\rho_i$ is equal to the sum of R_c and R_f.

The collision theory [15] and the concept of the limited flocculation process were originally used to calculate the values of R_c and R_f, respectively. Later, the diffusion theory [17, 18] was adopted by Fitch to calculate the value of R_c [31]. It was pointed out that the concentration of oligomeric radicals with chain length of j in the continuous aqueous phase, which was required to carry out the calculation, was very difficult to be determined.

3.2.2 Hansen–Ugelstad–Fitch–Tsai (HUFT) Model

To get around the above-mentioned problem, Hansen and Ugelstad [19, 32] proposed that a primary particle was nucleated when the chain length of an oligomeric radical reached a critical value (n^*) and then developed a mathematical model for calculating the number of primary particles per unit volume of water originating from homogeneous nucleation (N_p). This model begins with the following governing equation:

$$dN_p/dt = k_p[\mathrm{M}]_w([\mathrm{R}^*_{I,n^*}]_w + [\mathrm{R}^*_{M,n^*}]_w) \tag{3.17}$$

where $[\mathrm{M}]_w$ is the concentration of monomer in water, $[\mathrm{R}^*_{I,n^*}]_w$ is the concentration of oligomeric radicals with the critical chain length n^* originating from initiator radicals in the continuous aqueous phase, and $[\mathrm{R}^*_{M,n^*}]_w$ is the concentration of oligomeric radicals with the critical chain length n^* originating from the desorbed monomeric radicals in the continuous aqueous phase. Equation (3.17) implies that an oligomeric radical with the critical chain length (R^*_{I,n^*} or R^*_{M,n^*}) will readily coil up and then precipitate out of the continuous aqueous phase to form a particle nucleus immediately after the propagation reaction of this free radical with one more monomer molecule. The following summarizes some major assumptions made in the model development in order to obtain an analytic equation.

(a) Pseudo-steady-state assumption is applicable to the concentrations of free radicals in the continuous aqueous phase. In other words, changes in the concentrations of free radicals in water with time are insignificant.
(b) Contribution of oligomeric radicals originating from the desorbed monomeric radicals to particle nucleation is negligible.
(c) Absorption of free radicals by particle nuclei is neglected in the calculation of the total concentration of free radicals in the continuous aqueous phase, which is estimated to be $(\rho_i/k_{tw})^{1/2}$ in this case. The parameter k_{tw} is the bimolecular termination rate constant in water.

The resultant expressions for predicting the number of particle nuclei per unit volume of water as a function of time are as follows:

$$N_p(t) = (1/k_1)\left\{\left[k_1\rho_i n^* t + (k_2+1)^{n^*}\right]^{1/n^*} - (k_2+1)\right\} \tag{3.18}$$

$$k_1 = k_c/(k_p[\mathrm{M}]_w) \tag{3.19}$$

$$k_2 = (k_{tw}\rho_i)^{1/2}/(k_p[\mathrm{M}]_w) \tag{3.20}$$

where k_c is the average rate constant for the capture of free radicals by the particles.

Table 3.1. Solubility of a Variety of Monomers in Water and Other Related Parameters for Particle Nucleation

Monomer	Water Solubity (mmol dm^{-3})		$n^{*\,a}$	$C_0^{\,b}$ (mol dm^{-3})	$D_0^{\,b}$
Methyl acrylate	650	616			
Ethyl acrylate	150	184			
Propyl acrylate		44.7			
n-Butyl acrylate	11	10.9		1.5	0.9
n-Hexyl acrylate	1.2				
n-Octyl acrylate	0.34				
Methyl methacrylate		159	66	1.5	0.27
Ethyl methacrylate		45.4			
n-Butyl methacrylate		4.23			
Styrene	3.5		5	1.5	3.8
Vinyl chloride	170			1.5	0.25
Vinyl acetate	290		50	1.5	0.23
Reference	45	46	27, 57, 29	47, 48	47, 48

[a] Critical chain length of oligomeric radicals in the continuous aqueous phase.
[b] Parameters used to calculate the saturated concentration or the CMC of oligomers with chain length j in the continuous aqueous phase (C_j^*) in Eq. (3.48).

The reported values of the critical chain length of oligomeric radicals in the continuous aqueous phase for some common monomers in increasing order are styrene (5) [19] < vinyl acetate (50) [27] < methyl methacrylate (66) [29] (Table 3.1). The values in the parentheses represent the corresponding critical chain lengths of oligomeric radicals. It seems that the critical chain length of oligomeric radicals in the continuous aqueous phase is closely related to the solubility of monomers in water. The higher the solubility of monomer in water, the longer the critical chain length of oligomeric radicals in the continuous aqueous phase. As a consequence, the probability for relatively hydrophilic oligomeric radicals to enter monomer-swollen micelles to induce particle nucleation is greatly reduced. Furthermore, incorporation of a small amount of monomers containing carboxylic or hydroxyl groups such as acrylic acid, methacrylic acid, 2-hydroxyethyl acrylate, 2-hydroxyethyl methacrylate, and *N*-methylol acrylamide into the polymerization system prolongs the residence time of oligomeric radicals in the continuous aqueous phase and, therefore, promotes the formation of particle nuclei therein. In addition, the ionized carboxylic groups on the latex particle surfaces at pH 7–9 significantly enhance the electrostatic stabilization effect. This approach has been long put into practice in industry to manufacture latex products with improved colloidal stability and other key performance properties.

3.3 COAGULATIVE NUCLEATION

3.3.1 General Features of Coagulative Nucleation

While the homogeneous nucleation mechanism provides satisfactory theoretical groundwork for emulsion polymerization systems stabilized by surfactant at a level lower than its CMC, the general validity of this theory for those polymerization systems in the presence of monomer-swollen micelles is still open to debate. It is generally accepted that micellar nucleation predominates in emulsion polymerization of relatively hydrophobic monomers (e.g., styrene and butadiene) when the concentration of surfactant is greater than its CMC. On the other hand, homogeneous nucleation plays an important role in those polymerization systems with relatively hydrophilic monomers (e.g., methyl methacrylate, methyl acrylate and vinyl acetate) or in the absence of monomer-swollen micelles. Competition between micellar nucleation and homogeneous nucleation cannot be ruled out in emulsion polymerizations stabilized by surfactant at a level higher than its CMC.

In addition to micellar and homogeneous nucleation mechanisms, perhaps coagulative nucleation is another distinct mechanism that has been widely recognized in this field. Based on the positively skewed particle size distribution data obtained from the particle nucleation stage in emulsion polymerization of styrene stabilized by sodium dodecyl sulfate, Lichti et al. [33] and Feeney et al. [34, 35] proposed that the rate of formation of particle nuclei increased to a maximum and then decreased toward the end of the particle nucleation period (see Figure 3.4). This was attributed to the concept of coagulative nucleation (i.e., a two-step coagulative nucleation process). It was postulated that precursor particles were first generated by phase separation of

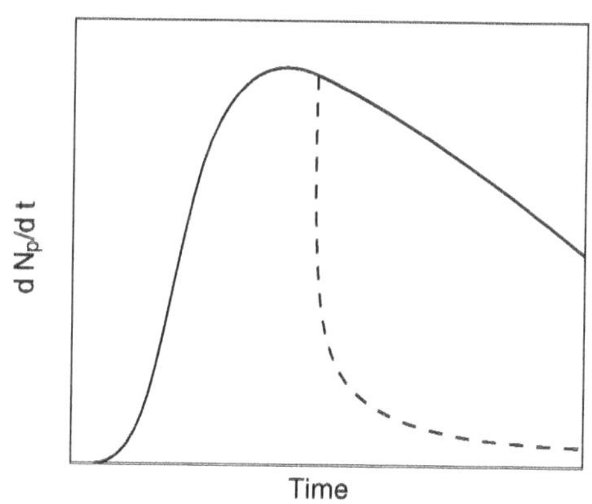

Figure 3.4. A schematic representation of the rate of particle nucleation as a function of time for the emulsion polymerization system that follows the coagulative nucleation mechanism. The solid and dashed lines represent the $d\,N_p/d\,t$ versus t profiles obtained from the emulsion polymerization systems without and with surfactant exhaustion, respectively.

oligomeric radicals out of the continuous aqueous phase. The term "precursor particles" is used to describe a primordial particle that is generated by the propagation of a single (or only a few) oligomeric radical species either in the continuous aqueous phase (homogeneous nucleation) or after entry of this radical into a monomer-swollen micelle (micellar nucleation). These precursor particles are extremely unstable in nature, and they tend to aggregate rapidly with one another in order to reduce the total oil–water interfacial area. The mature latex particles are produced through coagulation of the precursor particles rather than through growth of these precursor particles by polymerizing monomer therein. The number of mature latex particles per unit volume of water is primarily controlled by the amount of surfactant species available for stabilizing the particle nuclei. The Muller–Smoluchowski coagulation kinetics in combination with the DLVO theory was incorporated into the coagulative nucleation model. A mathematical model was then developed to predict the time evolution of the particle nucleation rate, the number of particle nuclei per unit volume of water, and the particle size distribution in emulsion polymerization.

3.3.2 Coagulative Nucleation Model Development

Feeney et al. [34] designated N_i ($i = 1, 2, 3, \ldots$) to be the number of particle nuclei produced by the limited flocculation of i precursor particles. Precursor particles are denoted by $i = 1$. The following equation was derived to calculate the rate of formation of particle nuclei comprising k precursor particles.

$$dN_k/dt = \left(\sum_{i=1}^{k-1} k_f N_i N_{k-i} - 2N_k \sum_{i=1}^{\infty} k_f N_i\right) + \delta_{k,1} g(t) \tag{3.21}$$

where $g(t)$ is the rate of generation of precursor particles through homogeneous nucleation and/or micellar nucleation processes. The value of $\delta_{k,1}$ is equal to zero for $k \neq 1$, whereas $\delta_{k,1}$ is equal to one for $k = 1$. Equation (3.21) shows that the rate of formation of particle nuclei comprising k precursor particles is equal to the rate of coagulation between the particle nuclei comprising i precursor particles and the particle nuclei comprising $k - i$ precursor particles minus the rate of coagulation between the particle nuclei comprising k precursor particles and the particle nuclei comprising i precursor particles ($i = 1, 2, 3, \ldots$) and plus the rate of generation of precursor particles through homogeneous nucleation and/or micellar nucleation processes. The parameter k_f is the particle coagulation rate constant, and it takes the following form based on the theory of Smoluchowski coagulation kinetics:

$$k_f = 4\pi D_w(2r)/W \tag{3.22}$$

where D_w is the diffusion coefficient of precursor particles in the continuous aqueous phase, r is the radius of precursor particles, and W is the Fuchs stability ratio. The total number of particles N ($=\Sigma N_k$, $k = 1, 2, 3, \ldots$) per unit volume of water can be written as

$$dN/dt = Q(t) - 4\pi D_w(2r)N_p^2/W \quad (3.23)$$

where $Q(t)$ is the rate of particle nucleation and it serves as an adjustable parameter in the computer simulation. With the assumption that the rate of particle nucleation is a time-independent function Q during the particle nucleation period, Feeney et al. [34] integrated Eq. (3.23) subject to the initial conditions ($N_i(t = 0) = 0$ for all i) to obtain the following expressions:

$$N(t) = \varepsilon_2 \tanh(\varepsilon_1 t/2) \quad (3.24)$$

$$\varepsilon_1 = 4[\pi D_w(2r)Q/W]^{1/2} \quad (3.25)$$

$$\varepsilon_2 = [WQ/(4\pi D_w(2r))]^{1/2} \quad (3.26)$$

In a similar manner, Eq. (3.21) also can be used to calculate the number of precursor particles per unit volume of water (N_1).

$$N_1(t) = (Q/\varepsilon_1)[\sinh(\varepsilon_1 t) + \varepsilon_1 t]/[\cosh(\varepsilon_1 t) + 1] \quad (3.27)$$

Assuming that a particle nucleus contains at least two precursor particles [34], the total number of particle nuclei (N_p) can be calculated as follows:

$$\begin{aligned} N_p(t) &= N - N_1 \\ &= \varepsilon_2 \tanh(\varepsilon_1 t/2) - (Q/\varepsilon_1)[\sinh(\varepsilon_1 t) + \varepsilon_1 t]/[\cosh(\varepsilon_1 t) + 1] \end{aligned} \quad (3.28)$$

In addition, the steady-state value of N_p at infinite time was given by

$$N_p(t \to \infty) = \varepsilon_2 - Q/\varepsilon_1 \quad (3.29)$$

Computer simulation results show a typical plot of the particle nucleation rate (dN_p/dt) as a function of time (t), as shown schematically in Figure 3.4. It is shown that the rate of particle nucleation increases to a maximum and then decreases toward the end of the particle formation period. The dashed line in Figure 3.4 is obtained by assuming the ceasing of particle nucleation at t_1 when the total particle surface area is equal to that covered by the surfactant molecules present in the polymerization system (i.e., $Q = 0$ when $t > t_1$). The time at which particle nucleation ceased can be estimated by the relationship $t_1 = 0.53\mu^{-2/5}\,[a_s S_0/(2fk_d\,[\mathrm{I}])]^{3/5}$ established in the Smith–Ewart model (see Section 3.1.1). Reasonable agreement between the model predictions with practical values of parameters and experimental data obtained from the emulsion polymerization of styrene was achieved.

It is noteworthy that both the homogeneous nucleation and coagulative nucleation mechanisms emphasize the significance of polymer reactions occurring initially in the continuous aqueous phase. Feeney et al. [36] used the small-angle neutron scattering technique to measure the average particle size of colloidal particles during the early stage of emulsion polymerization of styrene. The colloidal particles with an average diameter of 12 nm were observed, which was taken as evidence of the presence of precursor particles in the particle nucleation stage.

3.4 MIXED MODE OF PARTICLE NUCLEATION MECHANISMS

Hansen and Ugelstad [37–39] suggested that all the micellar nucleation, homogeneous nucleation and monomer droplet nucleation were operative in emulsion polymerization with a concentration of surfactant greater than its CMC. This indicates that monomer-swollen micelles and particle nuclei and emulsified monomer droplets compete with one another for the incoming oligomeric radicals from the continuous aqueous phase. Thus, the total rate of particle nucleation is given by

$$dN_p/dt = dN_m/dt + dN_h/dt + dN_d/dt \tag{3.30}$$

where the subscripts m, h, and d represent micellar nucleation, homogeneous nucleation, and monomer droplet nucleation, respectively. Assuming that the influence of desorption or reabsorption of free radicals is negligible, Eq. (3.30) can be written as

$$dN_p/dt = \rho_i(P_m + P_d + P_h) \tag{3.31}$$

where P_m and P_d are the probabilities of absorption of oligomeric radicals by monomer-swollen micelles and monomer droplets, respectively. P_h is the probability for the homogeneous nucleation events to take place in the continuous aqueous phase. The expressions proposed for P_m, P_d, and P_h are as follows.

$$P_k = 4\pi D_w r_k N_k F_k / (k_p[\mathrm{M}]_w + \textstyle\sum_k 4\pi D_w r_k N_k F_k) \tag{3.32}$$

$$P_h = k_p[\mathrm{M}]_w / (k_p[M]_w + \textstyle\sum_k 4\pi D_w r_k N_k F_k) \tag{3.33}$$

$$P_m + P_d + P_h + P_{p1} + P_{p0} = 1 \tag{3.34}$$

where the subscript k in Eqs. (3.32) and (3.33) denotes m, d, $p1$, or $p0$. P_{p1} and P_{p0} represent the probabilities of capture of oligomeric radicals by latex particles containing one free radical and zero free radical, respectively. The parameters r_k, N_k, and F_k are the radius, number, and reduction factor caused by the reversible diffusion and electrostatic repulsion of the water-borne oligomeric radicals of species k, respectively.

It should be noted that the assumption that only negatively charged oligomeric radicals with chain length of $n^* - 1$ or monomeric radicals can be captured by micelles, particle nuclei, and monomer droplets was adopted to derive Eqs. (3.32) and (3.33) [37]. The significance of this mechanism is the proposed competitive particle nucleation processes (i.e., micellar nucleation, homogeneous nucleation, and monomer droplet nucleation) involved in emulsion polymerization. Based on this mixed mode of particle nucleation mechanisms, several limiting cases including (a) homogeneous nucleation with potential absorption of oligomeric radicals by particle nuclei and negligible flocculation of particle nuclei [19], (b) limited flocculation of particle nuclei and negligible micellar nucleation and monomer droplet nucleation [19, 37], (c) competitive homogeneous nucleation and micellar nucleation with potential desorption and reabsorption of free radicals and negligible flocculation of particle nuclei [38], and (d) homogeneous nucleation and monomer droplet nucleation with negligible flocculation o f particle nuclei [39] were investigated.

Poehlein [40] summarized previous work and proposed a comprehensive particle nucleation mechanism involved in a persulfate initiated emulsion polymerization system, as shown schematically in Figure 3.5. Song and Poehlein [41, 42] developed a general kinetic model taking into account micellar nucleation, homogeneous nucleation, and monomer droplet nucleation in emulsion polymerization. The chain transfer and termination reactions occurring in the continuous aqueous phase, capture of oligomeric radicals by particle nuclei, and flocculation of particle nuclei were also incorporated into the model development. The resultant expressions for calculation of the rate of particle nucleation can be written as

$$
\begin{aligned}
(1/N_A)dN_p/dt &= d[\mathrm{P}]/dt \\
&= b\rho_i - B_c[\mathrm{P}] - B_f[\mathrm{P}]^2
\end{aligned}
\tag{3.35}
$$

where N_A is Avogadro's number, [P] is the concentration of particle nuclei, b is a parameter that takes into account the bimolecular termination reaction in the continuous aqueous phase, and B_c and B_f are kinetic parameters related to absorption of free radicals by particle nuclei and flocculation of particle nuclei, respectively. It should be noted that the effects of desorption and reabsorption of free radicals on particle nucleation are also incorporated into Eq. (3.35) through B_c. This is one of the major differences between Eqs. (3.35) and (3.16) developed by Fitch and Tsai [28–30]. This may result in a negative value of B_c and, as a result, the increased rate of particle nucleation with time (the characteristics of coagulative nucleation). It was reported that the curves of particle size distribution obtained just before the end of the particle nucleation stage always skewed toward small particle size, and this phenomenon could only be explained by the increased rate of particle nucleation with time [33–35].

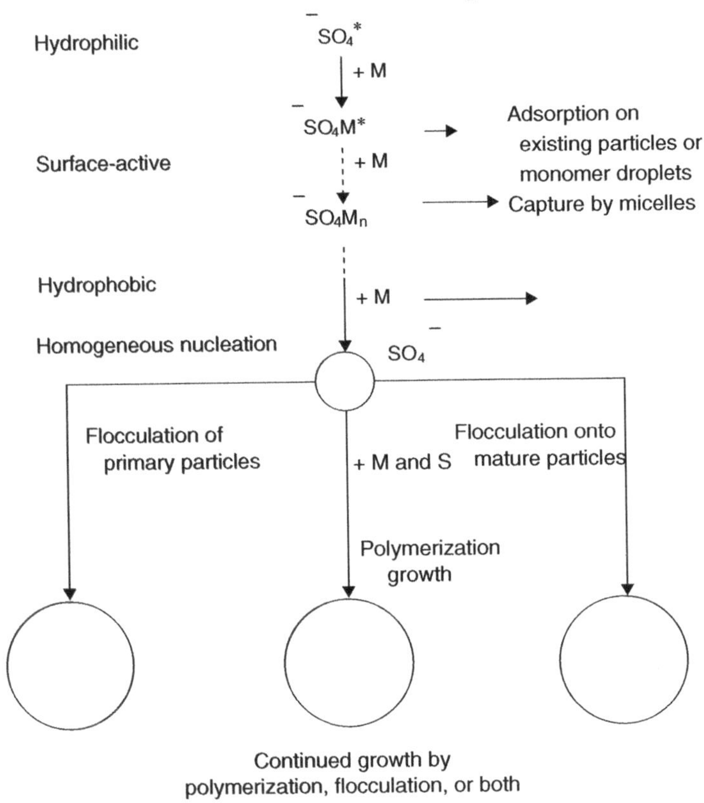

Figure 3.5. Paths for the formation of particle nuclei starting from persulfate initiator radicals generated in the continuous aqueous phase. The symbols M and S represent monomer and surfactant species, respectively.

Equation (3.35) can be used to simulate the number density of particle nuclei in the transient state. The analytic solution with two adjustable parameters (A_2 and τ) thus obtained is shown below:

$$N_p/N_{ps} = (1 - e^{-t/\tau})/[1 + (1/A_2)e^{-t/\tau}] \tag{3.36}$$

where N_{ps} is the number of particle nuclei per unit volume of water achieved at steady state. The values of N_{ps}, A_2, and τ can be estimated from the basic kinetic parameters and polymerization conditions. The validity of this comprehensive model was verified by the experimental data available in the literature (e.g., see Figure 3.6). The model accurately predicts the time evolution of N_p profile; the number of particle nuclei per unit volume of water first increases rapidly with time and then reaches a steady-state value (N_{ps}).

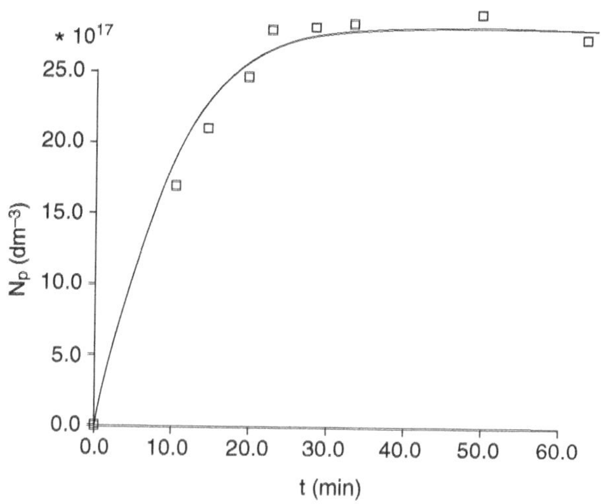

Figure 3.6. Time evolution of the number of particle nuclei in emulsion polymerization of styrene. The experimental data points are taken from the literature [44], and the solid line represents the model prediction with $A_2 = 1.188$, $\tau = 6.513\,\text{min}$, and $N_{ps} = 2.818 \times 10^{18}\,\text{dm}^{-3}$ [41, 42].

3.5 SURFACTANT-FREE EMULSION POLYMERIZATION

Surfactant-free emulsion polymerization is a unique process used to produce latex products with relatively large particle size and monodisperse particle size distribution with excellent water resistance and adhesion properties. The potential applications of these emulsion polymers include caulks and sealants, pressure-sensitive adhesives, and coatings. In the absence of surfactant, limited flocculation of latex particles greatly reduces the number of latex particles per unit volume of water (or increases the particle size) with the progress of polymerization. This will inevitably make the particle nucleation and growth mechanisms more complicated. Furthermore, intensive coagulation of the particles to form filterable solids and scraps adhering to the reactor wall and agitator could become a serious problem in plant production. Thus, the colloidal stability issue that has been sometimes ignored in the past must be addressed from both the theoretical and practical points of view. Incorporation of a small amount of functional monomers such as acrylic acid, methacrylic acid, or the polyethylene containing macromonomers that contribute to electrostatic or steric stabilization significantly improves the colloidal stability of the surfactant-free emulsion polymerization system. Micellar nucleation is generally ruled out for the polymerization system in the absence of monomer-swollen micelles.

A mechanistic model was developed to predict the number of latex particles per unit volume of water generated in surfactant-free emulsion polymerization

[43]. It was proposed that every oligomeric radical (or dead polymer chain) with chain length greater than $n^* - 1$ could form a rather stable particle embryo in the continuous aqueous phase. The mass balances on the active particle nucleus originating from the oligomeric radical (N_1) and inactive particle nucleus originating from the dead polymer chain (N_0) result in the following simultaneous differential equations:

$$dN_1/dt = k_p[\mathrm{M}]_w[R^*_{n^*-1}]_w N_A + 4\pi(r_0 N_0 - r_1 N_1)\sum_{j=1}^{n^*-1} D_{w,j}[R^*_j]_w N_A \quad (3.37)$$

$$dN_0/dt = (1-\lambda)(k_{tw}/2)\sum_{j=1}^{n^*-1}[R^*_{n^*-j}]w\sum_{i=j}^{n^*-1}[R^*_i]_w N_A + 4\pi(r_1 N_1 - r_0 N_0)\sum_{j=1}^{n^*} D_{w,j}[R^*_j]_w N_A \quad (3.38)$$

$$N_p = N_1 + N_0 \quad (3.39)$$

where $[\mathrm{R}^*_i]_w$ is the concentration of oligomeric radicals with chain length of i ($i = 1, 2, \ldots, n^* - 1$) in water, $D_{w,i}$ is the diffusion coefficient of oligomeric radicals with chain length of i ($i = 1, 2, \ldots, n^* - 1$) in water, r_0 and r_1 are the radii of the active particle nucleus originating from the oligomeric radical and inactive particle nucleus originating from the dead polymer chain, respectively, and λ is the ratio of the disproportionation termination reaction to the overall termination reaction. Derivation of Eqs. (3.37)–(3.39) is also based on the assumption that the bimolecular termination reaction taking place inside the particle nuclei is very fast.

The mass balances on each growing free radical in the continuous aqueous phase can be expressed as

$$d[R^*_1]_w/dt = 2fk_d[\mathrm{I}] - k_p[\mathrm{M}]_w[R^*_1]_w - k_{tw}[R^*_1]_w\sum_{j=1}^{n^*-1}[R^*_j]_w - 4\pi(r_1 N_1 + r_0 N_0)D_{w,1}[R^*_1]_w \quad (3.40)$$

$$d[R^*_j]_w/dt = k_p[\mathrm{M}]_w([R^*_{j-1}]_w - [R^*_j]_w) - k_{tw}[R^*_j]_w\sum_{j=1}^{n^*-1}[R^*_j]_w - 4\pi(r_1 N_1 + r_0 N_0)D_{w,j}[R^*_j]_w \qquad (j = 2, 3, \ldots n^* - 1) \quad (3.41)$$

Furthermore, the rate of changes in the total volume of active particle nuclei (V_1) or that of inactive particle nuclei (V_0) can be written as

$$dV_1/dt = k_p[\mathrm{M}]_w[R^*_{n^*-1}]_w n^* V' + k_p[\mathrm{M}]_p N_1 V'/N_A + 4\pi r_0 V_0\sum_{j=1}^{n^*-1} D_{w,j}[R^*_j]_w N_A + 4\pi r_0 N_0 V'\sum_{j=1}^{n^*-1} jD_{w,j}[R^*_j]_w N_A \quad (3.42)$$

$$dV_0/dt = (1-\lambda)(k_{tw}/2)\sum_{j=1}^{n^*-1}[R^*_{n^*-1}]_w\sum_{i=j}^{n^*-1}(n^* - j + i)[R^*_i]_w V' + 4\pi r_1 V_1\sum_{j=1}^{n^*-1} D_{w,j}[R^*_j]_w N_A + 4\pi r_1 N_1 V'\sum_{j=1}^{n^*-1} jD_{w,j}[R^*_j]_w - 4\pi r_0 V_0\sum_{j=1}^{n^*-1} D_{w,j}[R^*_j]_w N_A \quad (3.43)$$

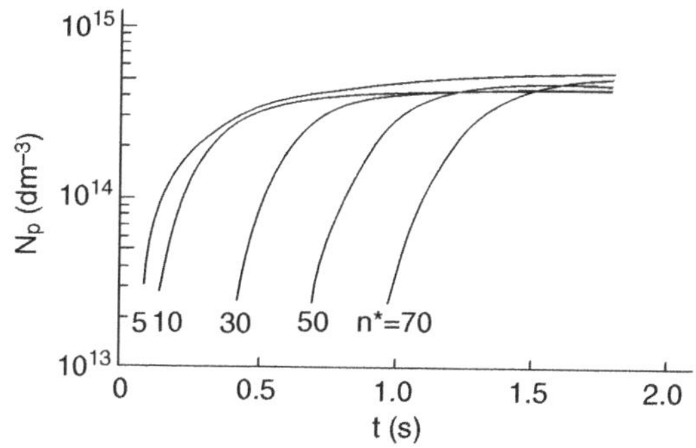

Figure 3.7. Number of particle nuclei per unit volume of water versus time profiles in emulsion polymerization at 50°C obtained from the model developed by Arai et al. [43]. The initial concentrations of monomer and initiator are 0.15 mol dm^{-3}-H_2O and 1×10^{-3} mol dm^{-3}, respectively.

$$V' = V_m \rho_m / [\rho_p / (1 - \Phi_m)] \tag{3.44}$$

where V_m is the molar volume of monomer, ρ_m and ρ_p are the densities of monomer and polymer, respectively, and Φ_m is the volume fraction of monomer in particle nuclei. The radii of active particle nuclei (r_1) and inactive particle nuclei (r_0) can be calculated by the following geometric relationships.

$$r_1 = [3V_1/(4\pi N_1)]^{1/3} \tag{3.45}$$

$$r_0 = [3V_0/(4\pi N_0)]^{1/3} \tag{3.46}$$

The simultaneous Eqs. (3.37)–(3.46) can be solved numerically, and typical computer simulation results are shown in Figure 3.7. For constant critical chain length of oligomeric radicals, the number of particle nuclei per unit volume of water increases with the progress of polymerization until a steady value is reached. In addition, the rate of particle nucleation increases with decreasing critical chain length of oligomeric radicals. This trend signifies the importance of the monomer solubility in water in the particle nucleation process. It is also interesting to note in Figure 3.7 that the ultimate number of particle nuclei per unit volume of water seems to be insensitive to changes in critical chain length of oligomeric radicals (i.e., monomer polarity). This is not consistent with one's experience that emulsion polymerization of relatively hydrophilic monomer(s) such as methyl methacrylate requires less stabilizer species (sulfate end-groups of polymer chains in this case) to maintain adequate colloidal stability during the reaction in comparison with polymerization of relatively hydrophobic monomer such as styrene. Furthermore, flocculation of particle nuclei to significantly reduce the oil–water interfacial free energy is a thermodynamically favorable process and it cannot be ignored in surfactant-free emulsion polymerization. These factors were not taken into account in the particle nucleation model of Arai et al. [43]. The data of water solubility of a variety of common monomers are summarized in Table 3.1. Depending on the solubility of monomer in water, the particle nucleation mechanisms involved in surfactant-free emulsion polymerization may be quite different.

Song and Poehlein [47, 48] developed a mathematical model to predict the rate of particle nucleation in surfactant-free emulsion polymerization. Precursor particles generated in the continuous aqueous phase are very unstable due to their small particle size and relatively low particle surface charge density. Therefore, they will flocculate with one another to form larger particle nuclei with sufficient particle surface charge density (i.e., enhanced colloidal stability). As a result, particle embryos originating from homogeneous nucleation grow primarily through particle flocculation. As soon as oligomeric radicals with critical chain length precipitate out of the continuous aqueous phase, particle embryos comprising m units of such oligomeric radicals form in the heterogeneous reaction system. Thus, the rate of particle nucleation can be expressed as

$$d[\mathrm{P}]/dt = k_p[\mathrm{M}]_w[R^*_{n^*-1}]_w/m - k_f[\mathrm{P}]^2 \qquad (3.47)$$

The first term on the right-hand side of Eq. (3.47) represents the rate of particle nucleation via homogeneous nucleation and/or *in situ* micellization of oligomeric radicals with critical chain length. The model development also took into consideration mole balances on various species in the continuous aqueous phase (oligomers with chain length j, oligomeric radicals with chain length j, and total free radicals) in order to calculate the first term on the right-hand side of Eq. (3.47). The saturated concentration or the CMC of oligomers with chain length j in the continuous aqueous phase (C^*_j) was assumed to decrease with increasing j^* according to the following empirical relationship that was originally developed for surfactants [49].

$$C^*_j = C_0 \exp(-D_0 j) \qquad (3.48)$$

where C_0 and D_0 are adjustable parameters (Table 3.1). The analytical solution to Eq. (3.47) is shown below.

$$N_p = \{N_A\rho/[m(d_0+1)\alpha_p E_1^{d_0}]\}t^{d_0+1} \qquad (3.49)$$

where ρ is the rate of generation of free radicals in the continuous aqueous phase that takes into account the effect of chain transfer reactions and α_p is the probability for free radicals to add one monomeric unit by the propagation

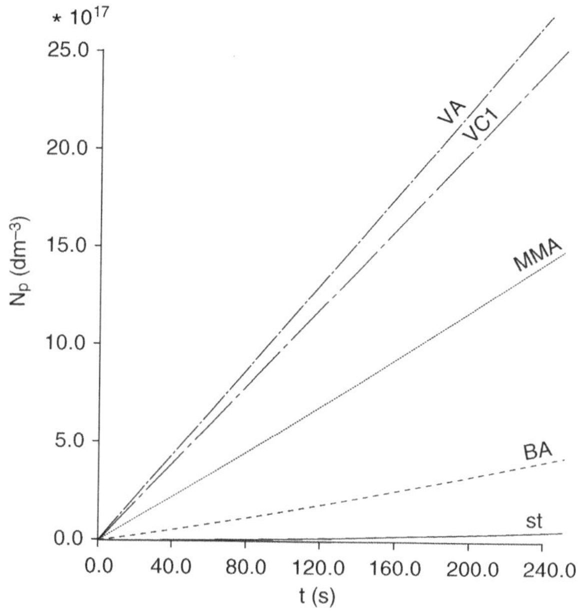

Figure 3.8. Number of particle nuclei per unit volume of water versus time profiles for surfactant-free emulsion polymerization of different monomers obtained from the model developed by Song and Poehlein [47, 48]. The symbols St, BA, MMA, VCl, and VA denote styrene, *n*-butyl acrylate, methyl methacrylate, vinyl chloride, and vinyl acetate, respectively.

reaction in the continuous aqueous phase. Parameters ρ, α_p, E_1 and d_0 are defined as follows:

$$\rho = \rho_i + \sum_i k_{tr,i}[R^*]_w[\mathrm{CTA}_i]_w \tag{3.50}$$

$$\alpha_p = k_p[\mathrm{M}]_w / \{k_p[\mathrm{M}]_w + k_{tw}[R^*]_w + k_c'[\mathrm{P}] + \sum_i k_{tr,i}[\mathrm{CTA}_i]_w\} \tag{3.51}$$

$$E_1 = C_0 k_p[\mathrm{M}]_w / \{\rho(k_{tw}[R^*]_w + \sum_i k_{tr,i}[\mathrm{CTA}_i]_w)\} \tag{3.52}$$

$$d_0 = -(\ln\alpha_p)/(D_0 + \ln\alpha_p) \tag{3.53}$$

where $[\mathrm{CTA}_i]_w$ and $k_{tr,i}$ are the concentration of chain transfer agent i in water and the corresponding chain transfer rate constant, respectively, and k_c' is the net rate coefficient for capture of free radicals by particle nuclei that includes the effect of desorption of free radicals out of the particle nuclei.

Representative computer simulation results for surfactant-free emulsion polymerizations of different monomers obtained from the model of Song and Poehlein [47, 48] are illustrated in Figure 3.8. The rate of particle nucleation during the early stage of polymerization in increasing order is styrene < *n*-butyl acrylate < methyl methacrylate < vinyl chloride < vinyl acetate. This trend correlates quite well with their water solubilities (see Table 3.1). The higher the water solubility of monomer, the faster the rate of particle nucleation. However, a larger population of particle nuclei generated during the early stage of particle nucleation does not necessarily guarantee a higher concentration of latex particles at steady state. For example, surfactant-free emulsion polymerization of relatively hydrophobic monomer (e.g., styrene) results in a slower particle nucleation rate but a longer particle nucleation period, thereby possibly achieving a higher concentration of latex particles at steady state. In addition to polymer reactions occurring in the continuous aqueous phase, the ultimate number of latex particles per unit volume of water is also controlled by other parameters such as polymer polarity and particle surface charge density.

3.6 EXPERIMENTAL WORK ON PARTICLE NUCLEATION

3.6.1 A Dilemma about Particle Nucleation Mechanisms

As discussed above, the well-known Smith–Ewart theory predicts that the number of particle nuclei per unit volume of water generated at the end of Interval I ($N_{p,\mathrm{I}}$) is proportional to the 0.6 power of the surfactant concentration and to the 0.4 power of the initiator concentration [Eq. (3.3)]. Accordingly, the rate of polymerization in Interval II ($R_p \sim N_{p,\mathrm{I}}$) is expected to behave in an identical fashion. These predictions very often form the basis of a test used to verify the validity of the Smith–Ewart theory in emulsion polymerization. The early experimental results mostly obtained from emulsion polymerization of styrene were reported to be in reasonable agreement with the theory under adequate conditions. However, deviations between the experimental data and the Smith–Ewart theory were also observed [13, 50, 51].

To determine which particle nucleation mechanism (micellar nucleation, homogeneous nucleation or coagulative nucleation) predominates in a particular emulsion polymerization system is not straightforward. For example, Roe [26] showed that even the fact that the experimental data obey the relationship $N_{p,\mathrm{I}} \sim S_0^{0.6}[\mathrm{I}]^{0.4}$ does not necessarily confirm the Smith–Ewart theory. This derivation is simply based on the assumptions that (a) flocculation of particle nuclei does not occur, (b) particle nucleation stops immediately after the depletion of surfactant available for stabilization of particle nuclei, and (c) the rate of generation of free radicals in the continuous aqueous phase is uniform. Thus, Eq. (3.3) alone cannot be used to distinguish between competitive particle nucleation mechanisms. More independent experimental data are required to distinguish micellar nucleation from homogeneous nucleation for emulsion polymerizations with the surfactant concentration above the CMC. Particle nucleation still represents an area of great challenge to polymer chemists, and it deserves more research efforts.

3.6.2 Some Representative Experimental Data of Particle Nucleation

The data of the number of particle nuclei per unit volume of water as a function of time may be useful in evaluating the validity of different particle nucleation mechanisms. Time evolution of the number of particle nuclei per unit volume of water reported in the literature generally fall into two major categories. The first type is that the number of particle nuclei per unit volume of water first increases and then reaches a constant value with the progress of polymerization. The emulsion polymerization systems that result in the first type of particle nucleation behavior include styrene [52], styrene with a surface-active anionic comonomer (sodium undecylenic isetheionate) [53], and vinyl acetate [54]. The second type of time evolution profile involves a rapidly increased number of particle nuclei per unit volume of water at the beginning of polymerization. This is followed by a decreased particle concentration and then the approach of a steady value during the particle nucleation stage. Examples are emulsion polymerization of vinyl acetate [55] and methyl methacrylate [28]. Recall that micellar nucleation plays an important role in emulsion polymerization of relatively hydrophobic monomers (e.g., styrene) when the concentration of surfactant is greater than its CMC. On the other hand, homogeneous nucleation has a significant influence on those polymerization systems with relatively hydrophilic monomers (e.g., methyl methacrylate and vinyl acetate) or in the absence of monomer-swollen micelles. Nevertheless, these experimental results do not exhibit any correlation with the water solubility of monomers (the water solubility of monomer in increasing order is styrene < methyl methacrylate < vinyl acetate, as shown in Table 3.1).

In addition to the water solubility of monomers, the characteristic of the N_p versus time curve is strongly dependent on other factors such as the amounts of surfactant molecules, sulfate end-groups of oligomeric radicals and anionic comonomer molecules (if present) available for stabilizing particle nuclei, the rate of generation of free radicals in the continuous aqueous phase, the ionic strength of aqueous solution and the agitation speed. For example, emulsion polymerization of styrene normally results in a much smaller population of particle nuclei, as illustrated in Figure 3.8. The total surface area of these particle nuclei is not very large and, therefore, the colloidal stability is sufficient for preventing the particle nuclei from intensive flocculation. The presence of a small quantity of anionic comonomer further increases the particle surface charge density and, thus, enhances the colloidal stability of the styrene emulsion polymerization system. As a result, the first type of the N_p versus t profile is achieved [52, 53]. Although the number of particle nuclei per unit volume of water generated is much larger (Figure 3.8), emulsion polymerization of relatively hydrophilic vinyl acetate may also exhibit the first type of particle nucleation [54]. This is because the emulsion polymerization system is adequately stabilized and, hence, the influence of the limited particle flocculation is insignificant. On the other hand, the rate of particle nucleation is much faster

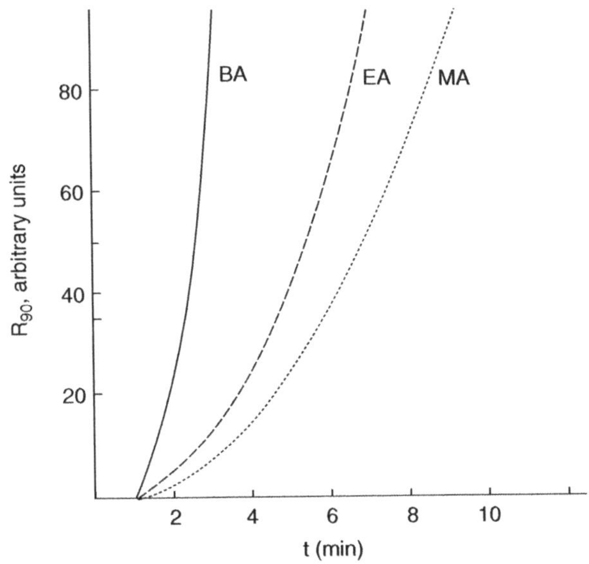

Figure 3.9. Rayleigh light scattering intensity versus time data obtained from the surfactant-free emulsion polymerizations of different monomers. The symbols MA, EA, and BA denote methyl acrylate, ethyl acrylate, and *n*-butyl acrylate, respectively.

in emulsion polymerizations of relatively hydrophilic monomers such as vinyl acetate and methyl methacrylate as compared to the styrene counterpart (Figure 3.8). Judging from the extremely large oil–water interfacial area and the rather low particle surface charge density (i.e., the very high oil–water interfacial free energy) associated with emulsion polymerization of vinyl acetate or methyl methacrylate, the probability for the limited flocculation of particle nuclei to take place increases significantly with decreasing stabilization effect. This is one of the major reasons why the second type of particle nucleation is very often observed in emulsion polymerizations of relatively hydrophilic monomers such as vinyl acetate [55] and methyl methacrylate [28].

Rayleigh light scattering intensity (R_{90}) data as a function of time for surfactant-free emulsion polymerizations of different acrylic monomers are shown in Figure 3.9 [56]. The rates of change in R_{90} with time in increasing order is methyl acrylate < ethyl acrylate < *n*-butyl acrylate. Rayleigh light scattering intensity is proportional to the concentration of particle nuclei and the volume of particle nuclei to the first power and the second power, respectively. As would be expected, the rate of particle nucleation or the concentration of particle nuclei in increasing order is n-butyl acrylate < ethyl acrylate < methyl acrylate (Table 3.1). In other words, the resultant average volume of particle nuclei in increasing order is methyl acrylate < ethyl acrylate < *n*-butyl acrylate.

Moreover, the effect of the volume of particle nuclei on the Rayleigh light scattering intensity overrides that of the concentration of particle nuclei, and this is reflected in the R_{90} versus time profiles shown in Figure 3.9. Again, these experimental data demonstrate the important role of polarity of monomers in the particle nucleation period, and they are consistent with the theoretical work of Song and Poehlein [47, 48].

In summary, the rate of particle nucleation and the ultimate number of particle nuclei per unit volume of water formed in emulsion polymerization are closely related to the water solubility of monomers, the rate of generation of free radicals in the continuous aqueous phase and the amounts of surfactants and other stabilizing species (e.g., comonomers containing sulfate or sulfonate groups and ionized carboxyl groups and protective colloids) available for stabilizing the particle embryos. Based on the levels of surfactants and other stabilizers present in the emulsion polymerization system, the existing particle nuclei tend to adjust their sizes (i.e., to decrease the total particle surface area) through limited particle flocculation in order to reduce the interfacial free energy. This particle flocculation process is thermodynamically feasible.

Aqueous polymerization of methyl methacrylate in the absence of both monomer-swollen micelles and emulsified monomer droplets was carried out to study the role of oligomeric radicals generated in the continuous aqueous phase [29]. Vapor pressure osmometry and gel permeation chromatography were employed to characterize the oligomers formed in the continuous aqueous phase. The maximum degree of polymerization of the growing methyl methacrylate chain with a sulfate or sulfonate group originating from the initiator is in the range of 65–75 units. This result suggests that the critical chain length of oligomeric radicals (n^*) for the methyl methacrylate emulsion polymerization system is about 70. A value of n^* equal to 50 was reported for emulsion polymerization of vinyl acetate [27]. A value of n^* equal to 5 was used in the mathematical modeling of emulsion polymerization of styrene [57]. These values of n^* are also included in Table 3.1. The critical chain length of oligomeric radicals in the continuous aqueous phase is an important parameter in the homogeneous nucleation mechanism, and more research efforts are required to determine the values of n^* for emulsion polymerization systems with other monomers.

Wang and Poehlein [58, 59] and Thomson et al. [60] isolated and characterized the water-soluble oligomers produced during emulsion polymerization. Analytical techniques such as FTIR spectroscopy, mass spectroscopy, and ^{13}C-NMR were used to characterize these oligomers. Wang and Pan [61] studied the surfactant-free emulsion polymerization of styrene with a water-soluble comonomer 4-vinylpyridine. At the very beginning of copolymerization of styrene and 4-vinylpyridine in the continuous aqueous phase, oligomers rich in the monomeric unit of 4-vinylpyridine are generated. It was postulated that these surface-active oligomers form monomer-swollen micelles that are available for the subsequent particle nucleation. In addition, they are capable of stabilizing monomer droplets. This is followed by the typical particle growth period. Ou et al. [62] investigated the effect of addition of a hydrophilic comonomer (vinyl acetate or methyl methacrylate) on particle nucleation in the surfactant-free emulsion polymerization of styrene. The gel permeation chromatography data clearly show a population of oligomers with molecular weight of about 1000 g mol^{-1} obtained from the early stage of polymerization. This is attributed to the micellar nucleation mechanism. It is noteworthy that emulsion copolymerization of styrene with vinyl acetate is not feasible due to their unfavorable reactivity ratios. As a matter of fact, severe poisoning of the emulsion polymerization of vinyl acetate may be experienced if the reaction system were contaminated by styrene. Isolation of the above-mentioned oligomers and characterization of their surface activity in the continuous aqueous phase are of great interest to gaining a fundamental understanding of the related particle nucleation mechanisms. These studies provide valuable information on the nature and of oligomeric radicals formed early in the emulsion polymerization and promote the fundamental understanding of the particle nucleation process.

Yan et al. [63] investigated the surfactant-free emulsion copolymerization of styrene, methyl methacrylate, and acrylic acid initiated by ammonium persulfate. As expected, both the rates of particle nucleation and polymerization increase with increasing concentration of acrylic acid or initiator. The persulfate initiator concentration and polymerization temperature are the predominant parameters that govern the particle nucleation process (homogeneous nucleation accompanied with limited flocculation). A shell growth mechanism was proposed to describe the particle growth stage. Mahdavian and Abdollahi [64] carried out the surfactant-free emulsion copolymerization of styrene and butadiene in the presence of various levels of acrylic acid. The minor comonomer, acrylic acid, has a significant influence on particle nucleation. Both the number of latex particles per unit volume of water and the rate of polymerization increase with increasing concentration of acrylic acid. However, there is no significant difference in the rate of polymerization per particle in all the polymerizations investigated. Furthermore, the growth of latex particles is less sensitive to changes in the concentration of acrylic acid. Zhang et al. [65] prepared cationic emulsion copolymers of styrene, *n*-butyl acrylate, and *N,N*-dimethyl,*N*-butyl,*N*-methacryloloxylethyl ammonium bromide via the surfactant-free emulsion polymerization process. Azobis(isobutyramidine hydrochloride) was used as a cationic initiator. Methanol was employed to improve the solubility of monomers in the continuous aqueous phase. The latex particle size decreased with increasing concentration of cationic comonomer or initiator. By contrast, the latex particle size first decreased and then increased with increasing concentration of methanol. It was postulated that particle nuclei are generated via both the micellar nucleation and homogeneous nucleation mechanisms based on the latex particle size and polymer molecular weight data.

Sahoo and Mohapatra [66] studied the catalytic effect of the *in situ* developed Cu(II)-EDTA complex with ammonium persulfate on the surfactant-free emulsion polymerization of methyl methacrylate. The rate of polymerization at 50 °C is proportional to the concentrations of Cu(II), EDTA, ammonium persulfate, and methyl methacrylate to the 0.35, 0.69, 0.57, and 0.75 powers, respectively. In addition, the apparent activation energy and activation energies of the initiator decomposition, propagation, and termination reactions, respectively, are 34.5, 26.9, 29, and 16 kJ mol^{-1}. It was proposed that the complex just acts as an effective surfactant in stabilizing the polymethyl methacrylate nanoparticles nucleated during polymerization. Independent experiments are required to verify this speculation and clarify the related stabilization mechanism.

As aforementioned, micellar nucleation is generally not considered as an appropriate mechanism for the formation of latex particles in surfactant-free emulsion polymerization. This point of view has been reconfirmed [63, 67]. However, in recent studies on surfactant-free emulsion polymerizations [61, 62, 65], it has been postulated that surface-active oligomers form *in situ* and then aggregate together to form monomer-swollen micelles in the continuous aqueous phase during the early stage of polymerization. The polymerization in the presence of functional monomers is especially prone to follow this particle nucleation mechanism. How to reconcile the particle nucleation mechanisms in dispute represents a great challenge to colloid and polymer scientists.

Sutterlin [46] studied the effect of the polarity of various monomers (styrene, acrylate ester monomers, and methacrylate ester monomers; see Table 3.1) on the particle nucleation mechanisms involved in emulsion polymerization. When the surfactant concentration is above its CMC, the emulsion polymerization of styrene follows the Smith–Ewart theory ($N_{p,\mathrm{I}} \sim S_0^{0.6}$) except those experiments with relatively low levels of surfactant. The exponent x in the relationship $N_{p,\mathrm{I}} \sim S_0^x$ decreases with increasing monomer polarity when the surfactant concentration is above its CMC. This trend is attributed to the increased tendency of agglomeration of particle nuclei with monomer polarity. The emulsion polymerizations of less polar monomers deviate significantly from the Smith–Ewart theory ($x >> 0.6$) if the surfactant concentration is reduced to a level just below its CMC. This implies that some mechanisms other than micellar nucleation (homogeneous nucleation or coagulative nucleation) must operate in these emulsion polymerization systems.

Varela de la Rosa et al. [68–70] carried out emulsion polymerizations of styrene stabilized by sodium dodecyl sulfate and initiated by potassium persulfate at 50 °C. They proposed the following reaction mechanism to describe the conventional styrene emulsion polymerization system.

(a) Both the number of particle nuclei per unit volume of water and the rate of polymerization increase throughout the particle nucleation stage. Micellar nucleation is the predominant mechanism, and monomer-swollen micelles disappear in the monomer conversion range of 5–10%, which marks the end of Stage I.

(b) In Stage II, both the number of particle nuclei per unit volume of water and the rate of polymerization increase continuously with increasing conversion, but at slower rates. Particle nuclei form via homogeneous nucleation, provided that emulsified monomer droplets and sufficient surfactant ($>5 \times 10^{-5}$ M) are present in the polymerization system. This stage ends immediately after the depletion of emulsified monomer droplets. However, particle nucleation does not necessarily cease at this point of time.

(c) The number of latex particles per unit volume of water produced during polymerization is proportional to the 0.36 power of the initiator concentration.

Recently, Herrera-Ordonez et al. [71, 72] developed a mechanistic model to study the particle nucleation mechanisms involved in the emulsion polymerization of styrene stabilized by sodium dodecyl sulfate at a level greater than its CMC. It was shown that micellar nucleation governs the particle nucleation process and flocculation of particle nuclei is insignificant. Furthermore, the number of particle nuclei originating from micellar nucleation is at least 10 orders of magnitude greater than that of particle nuclei stemming from homogeneous nucleation, even in the emulsion polymerization of relatively hydrophilic methyl methacrylate [73].

3.6.3 Some Potential Techniques for Studying Particle Nucleation

Nomura et al. [74, 75] proposed an experimental method to study the competitive particle nucleation mechanisms (micellar nucleation versus homogeneous nucleation) in a given emulsion polymerization system. This approach involves the emulsion copolymerization of relatively hydrophobic styrene with relatively hydrophilic monomers such as methyl methacrylate or methyl acrylate. The composition of copolymer produced during the very early stage of polymerization (far lower than 1% monomer conversion), which reflects the characteristic of copolymer at the locus of particle nucleation, is then determined. Emulsion copolymerization of styrene with methyl methacrylate (or methyl acrylate) was carried out, where sodium dodecyl sulfate was used to stabilize the emulsion polymerization system and where the weight ratio of styrene to methyl methacrylate (or methyl acrylate) was kept constant at 1:1. The experimental results show that the compositions of copolymers obtained from emulsion polymerizations in the presence and absence of monomer-swollen micelles are quite different. This provides supporting evidence of the generally accepted Smith–Ewart theory that micellar nucleation controls the particle nucleation process in the emulsion copolymerization of styrene with methyl methacrylate (or methyl acrylate) in the presence of monomer-swollen micelles.

In the surfactant-free emulsion polymerization of styrene, Feeney et al. [36] used the small-angle neutron scattering (SANS) technique in combination with an aqueous polyacrylamide gel containing initiator to measure the colloidal particle size during the early stage of polymerization. The presence of particle nuclei (i.e., precursor particles) with an average radius of 6nm was observed. It was demonstrated that SANS is a very effective technique for the investigation of particle nucleation mechanisms involved in emulsion polymerization. However, the colloidal system involved is much more complicated than conventional emulsion polymerization, and the influence of the aqueous polyacrylamide gel on the polymerization mechanisms should be evaluated cautiously. For example, flocculation of particle nuclei may become diffusion-controlled due to very high viscosity, and bridging flocculation of particle nuclei induced by water-soluble polyacrylamide chains with very high molecular weight may also occur during polymerization. In addition, the cage effect of initiator radicals may become more important, and the bimolecular termination reaction (if present) may also be retarded significantly due to the gel effect in the continuous aqueous phase. All these factors need to be taken into consideration in the interpretation of the experimental data. Kuhn and Tauer [76] developed a technique of on-line monitoring of the optical transmission and conductivity of the reaction mixture to study the particle nucleation process in the surfactant-free emulsion polymerization of styrene. It was concluded that the rate of initiation in the continuous aqueous phase plays an important role in the particle nucleation stage. Particle nucleation is induced via the cluster formation of oligomers in the continuous aqueous phase. Later, Tauer and Deckwer [77] used the MALDI-TOF-MAS technique to characterize the end-groups of polymer chains obtained from the surfactant-free emulsion polymerization of styrene initiated by potassium persulfate. Surprisingly, a variety of end-groups were identified in addition to the sulfate group originating from the persulfate initiator. It was then concluded that polymer chains started with oligomeric radicals generated by side reactions in the continuous aqueous phase play an important role in the particle nucleation process (e.g., homogeneous nucleation). Moreover, they pointed out that the surface activity of oligomeric radicals absorbed by the latex particles is definitely not a prerequisite. Recently, Kozempel et al. [78] used the on-line multi-angle laser light scattering technique to study the surfactant-free emulsion polymerization of styrene. It was proposed that the polymerization mechanisms are characterized by three distinct intervals. Monomer droplets (~200nm in diameter) are produced during Interval A. This is followed by the formation of particle nuclei (Interval B). Immediately after the start of particle nucleation, latex particles absorb monomer molecules from monomer droplets, thereby leading to the depletion of monomer droplets and a reduction in the average size of the scattering objects. Beyond Interval B, the average size of the scattering objects increases again as a result of the predominant growth of latex particles (Interval C). This is presumably due to the limited particle flocculation process.

Chern and Lin [79] used an extremely water-insoluble blue dye as the probe for determining the loci of particle nucleation in the emulsion polymerization of styrene. Measurements of the weight percentage of dye incorporated into the final latex particles provide valuable information on particle nucleation mechanisms. The rationale of this approach is that the amount of dye within the resultant polymer particles is negligible if homogeneous nucleation governs the particle formation process. This is because transport of dye molecules from monomer-swollen micelles to particle nuclei is prohibited due to the extremely hydrophobic nature of dye. On the other hand, the amount of dye within the resultant polymer particles is comparable to that of dye originally solubilized in the monomer-swollen micelles provided that particle nuclei are primarily generated by micellar nucleation. This work illustrates that micellar nucleation and homogeneous nucleation compete with each other when the concentration of surfactant is higher than its CMC. In contrast, most of the particle nuclei are generated via homogeneous nucleation in the emulsion polymerization of styrene in the absence of monomer-swollen micelles. This dye technique was then applied to the emulsion polymerization of more polar methyl methacrylate [80]. The water solubility of methyl methacrylate is about 80 times as that of styrene. The polymerization taking place in the continuous aqueous phase—and, hence, homogeneous nucleation—is greatly enhanced in the emulsion polymerization of methyl methacrylate compared to the styrene counterpart. Indeed, the experimental data clearly show that homogeneous nucleation plays a key role in the early stage of the emulsion polymerization of methyl methacrylate. A mixed mode of particle nucleation (micellar nucleation and homogeneous nucleation) is operative in the polymerization system when the concentration of surfactant is higher than the CMC. On the other hand, homogeneous nucleation is the predominant mechanism that governed the population of particle nuclei produced if there were no monomer-swollen micelles present in the polymerization system.

Rudschuck et al. [81] adopted (on-line) fluorescence spectroscopy to investigate the nucleation and growth of particle nuclei during the surfactant-free emulsion polymerization of styrene. The polymerization is initiated by a macroinitiator, the hydrolyzed propene–maleic acid copolymer with *t*-butyl perester groups. Pyrene molecules are incorporated into the backbone of the macroinitiator to probe the polymerization mechanisms. Four distinct regions are identified during polymerization. The first stage is simply related to the initial heating period. The particle nucleation process begins with thermal decomposition of the perester groups into *t*-butyl-hydroxyl radicals, aliphatic radicals at the chain and carbon dioxide. The free radicals attached to the macroinitiator backbone polymerize with styrene molecules dissolved in the continuous aqueous phase. The resultant macroinitiator species with grafted oligostyrene chains exhibits some surface activity and contributes to the stabilization of polystyrene particle nuclei generated in the continuous aqueous phase. Adsorption of the surface-active macroinitiator is reflected in the decreased fluorescence intensity ratio I_1/I_3 as well as the increased fluorescence intensity and the back-scattered light due to the formation of particle nuclei. The fluorescence intensity ratio I_1/I_3 is defined as the ratio of the peak height of the first vibronic band to the third vibronic band of the emission spectra of pyrene. It represents a quantitative measure of the nonpolar nature of microenvironment in which most of the hydrophobic pyrene molecules reside. This is followed by the relatively constant fluorescence intensity ratio I_1/I_3 and the increased fluorescence intensity. The former is attributed to the attachment of macroinitiator onto the particle surface. As for the latter, it is caused by the growth of latex particles at the expense of monomer droplets. Finally, the fluorescence intensity increases rapidly toward the end of polymerization as a result of the gel effect.

Tauer et al. [82] demonstrated that the heat flow arising from free radical polymerization versus time (or monomer conversion) profiles obtained from reaction calorimetry clearly reflect changes in recipe ingredients and experimental conditions in emulsion polymerizations. Varela De La Rosa et al. [83] then used the reaction calorimetry technique to illustrate the significant implication of the maximum rate of polymerization in the emulsion polymerization of styrene. It was proposed that the point at which the maximum rate of polymerization occurs signifies the end of the particle nucleation stage and the disappearance of monomer droplets. This technique was also used to study the effect of the initial ratio of monomer to water on the emulsion polymerization of styrene [84]. The surfactant concentration (40 mM sodium dodecyl sulfate) was kept at a constant level higher than the CMC. The experimental data show that, at low ratios of monomer to water (final total solids content ≦10%), micellar nucleation takes place throughout the polymerization and the constant polymerization rate period is not achieved. On the other hand, at high ratios of monomer to water (final total solids content ≥20%), particle nuclei form first via the predominant micellar nucleation mechanism. This is followed presumably by homogeneous nucleation immediately after the depletion of monomer-swollen micelles. Furthermore, the length of the homogeneous nucleation period increases with increasing ratio of monomer to water. These particle nucleation phenomena are supported by the polymerization kinetic data. Based on these experimental results, it was proposed that the generally recognized Interval II could not be characterized by the constant number of latex particles per unit volume of water and the constant rate of polymerization for the reaction system with a surfactant concentration greater than its CMC. The characteristics of Interval II include the continuous particle nucleation process and the increased rate of polymerization. As expected, Interval II ends immediately after the disappearance of monomer droplets. Nevertheless, particle nucleation might not cease at this time. This study illustrates that the information on the evolution of particle nuclei throughout the polymerization is required to monitor the whole particle nucleation process.

Sajjadi [85] investigated the diffusion-controlled nucleation and growth of particle nuclei in the emulsion homopolymerizations of styrene and methyl methacrylate. The polymerization starts with two stratified layers of monomer and water containing surfactant and initiator, with the water layer being stirred gently. In this manner, the rate of transport of monomer becomes diffusion-limited. As a result, the rate of growth of particle nuclei is reduced significantly, and more latex particles can be nucleated in emulsion polymerization.

3.6.4 Effects of Surfactant Concentration on Particle Nucleation

Regardless of which particle nucleation mechanisms predominate in the particle formation process, the amount of surfactant available for stabilizing particle nuclei is perhaps the most important parameter that controls the size of the population of latex particles produced during emulsion polymerization. Figure 3.10 shows a schematic representation of the number of latex particles per unit volume of water as a function of the concentration of surfactant initially present in the polymerization system. For emulsion polymerizations carried out at surfactant concentrations lower than the CMC, the number of latex particles per unit volume of water first remains relatively constant and then increases with increasing surfactant concentration. This is followed by the rapidly increased number of latex particles per unit volume of water with surfactant concentration when the surfactant concentration is in the vicinity of the CMC. The number of latex particles per unit volume of water then levels off when the surfactant concentration is increased to a level well above the CMC.

At very low surfactant concentration, the number of latex particles per unit volume of water is relatively insensitive to changes in the surfactant concentration. This indicates that the initiation reactions taking place in the continuous aqueous phase are the primary mechanism responsible for the particle nucle-

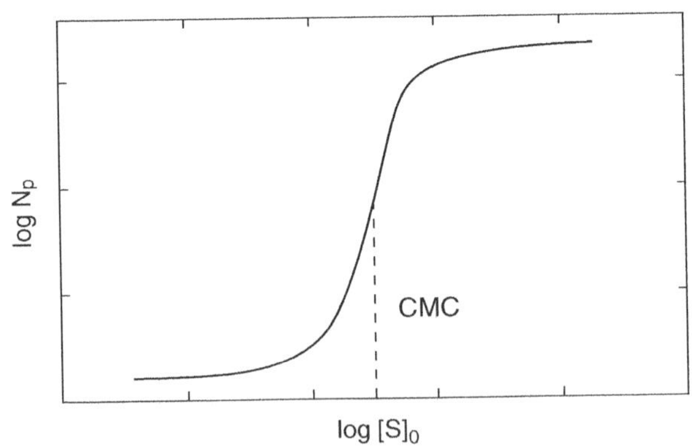

Figure 3.10. A schematic representation of the number of latex particles per unit volume of water as a function of the concentration of surfactant initially present in the polymerization system.

ation process. The colloidal stability of particle nuclei is achieved primarily by the electrostatic stabilization effect of the anionic sulfate end-groups of polymer chains originating from the persulfate initiator rather than by the adsorption of surfactant on the particle surfaces. As the surfactant concentration is increased, both the free surfactant molecules dissolved in water and negatively charged oligomeric radicals generated in the continuous aqueous phase synergistically contribute to the stabilization of particle nuclei. As a consequence, the number of latex particles per unit volume of water increases significantly with increasing surfactant concentration. When the surfactant concentration is higher than the CMC, the rapidly increased number of latex particles per unit volume of water with surfactant concentration can be attributed to the formation of monomer-swollen micelles that directly provide the loci for particle nucleation to occur therein or simply act as reservoirs to supply the existing or newly borne particle nuclei with surfactant molecules for adequate colloidal stability. Thus, the homogeneous nucleation mechanism cannot be ruled out. Furthermore, the number of latex particles per unit volume of water is primarily determined by the amount of surfactant initially present in the polymerization system, and the contribution of the anionic sulfate end-groups of polymer chains originating from the persulfate initiator to particle nucleation becomes insignificant. At very high surfactant concentration, the number of latex particles per unit volume of water is controlled by the rate of generation of free radicals in the continuous aqueous phase. Only a small fraction of monomer-swollen micelles may have a chance to capture free radicals from the continuous aqueous phase to induce particle nucleation therein. It is also possible that the total surface area of particle nuclei generated is not large enough to allow a significant fraction of surfactant molecules to adsorb on the particle surfaces.

It should be noted that the probability for the continuous particle nucleation throughout the emulsion polymerization increases with increasing surfactant concentration. For constant monomer weight, the higher the surfactant concentration, the smaller the latex particles produced in the emulsion polymerization system. In addition, the longer the particle nucleation period, the broader the residence time distribution of particle nuclei within the reactor (i.e., the broader the resultant particle size distribution). These rules of thumb, based on a large number of fundamental studies on nucleation and growth of particle nuclei, have been widely used in industry to effectively use surfactant to stabilize various latex products with balanced performance properties.

3.7 NONIONIC AND MIXED SURFACTANT SYSTEMS

Mixed anionic and nonionic surfactant systems have been widely used in industry to manufacture latex products. Anionic surfactants can provide electrostatic repulsion force between two similarly charged electric double layers. By contrast, nonionic surfactants can impart two approaching latex particles with the steric stabilization mechanism. In addition, nonionic surfactants can improve the chemical and freeze–thaw stability of latex products. The following studies clearly show that the reaction mechanisms and kinetics involved in a variety of emulsion polymerizations stabilized by nonionic or mixed anionic/nonionic surfactant systems are far more complicated than the conventional Smith-Ewart theory.

3.7.1 Nonionic Surfactant Systems

Ozdeer et al. [86] studied the emulsion copolymerization of styrene and *n*-butyl acrylate using octylphenol polyethoxylate with an average of 40 monomeric units of ethylene oxide per molecule (Triton X-405) as the sole nonionic surfactant. Although the levels of Triton X-405 are all well above its CMC, the substantial partitioning of Triton X-405 molecules into the oil phase leads to the scenario that the concentrations of Triton X-405 in the continuous aqueous phase are well below the CMC for emulsion polymerizations with the lowest to intermediate levels of Triton X-405. As a consequence, latex products with unimodal particle size distribution are achieved for emulsion polymerizations with the lowest and highest levels of Triton X-405. On the other hand, latex products with bimodal particle size distribution are obtained from emulsion polymerizations with intermediate levels of Triton X-405. These experimental results are attributed to homogeneous nucleation and coagulative nucleation for emulsion polymerizations with lower levels of Triton X-405, to homogeneous nucleation and coagulative nucleation followed by micellar nucleation for emulsion polymerizations with intermediate levels of Triton X-405, and then to micellar nucleation for emulsion polymerizations with higher levels of Triton X-405.

Capek and Chudej [87] studied the emulsion polymerization of styrene stabilized by polyethylene oxide sorbitan monolaurate with an average of 20 monomeric units of ethylene oxide per molecule (Tween 20) and initiated by the redox system of ammonium persulfate and sodium thiosulfite. It is interesting to note that the constant reaction rate period is not present in this polymerization system. The maximal rate of polymerization is proportional to the initiator and surfactant concentrations to the –0.45 and 1.5 powers, respectively. The final number of latex particles per unit volume of water is proportional to the initiator and surfactant concentrations to the 0.32 and 1.3 powers, respectively. In addition, the resultant polymer molecular weight is proportional to the initiator and surfactant concentrations to the 0.62 and –0.97 powers, respectively. Some possible reaction mechanisms may explain the deviation of the polymerization system from the classical Smith–Ewart theory. Lin et al. [88] investigated the emulsion polymerization of styrene stabilized by nonylphenol polyethoxylate with an average of 40 monomeric units of ethylene oxide per molecule (NP-40) and initiated by sodium persulfate. The rate of polymerization versus monomer conversion curves exhibit two nonstationary reaction rate intervals and a vague constant rate period in between.

The rate of polymerization and the final number of latex particles per unit volume of water are proportional to the 1.4 and 2.4 powers, respectively, of the NP-40 concentration. The polymerization system does not follow the conventional micellar nucleation model, and some possible reaction mechanisms responsible for this deviation are discussed.

Ouzineb et al. [89] carried out emulsion copolymerizations of *n*-butyl acrylate and methyl methacrylate with different types and concentrations of surfactants (Triton X-405 versus sodium dodecyl sulfate) to study particle nucleation and the resultant latex particle size and particle size distribution. The presence of relatively hydrophilic methyl methacrylate in the continuous aqueous phase has a significant influence on the CMC of Triton X-405. Furthermore, the relatively hydrophobic *n*-butyl acrylate predominates in the particle nucleation process involved in emulsion copolymerizations of *n*-butyl acrylate and methyl methacrylate, with the final number of latex particles per unit volume of water very similar to that of latex particles obtained from the homopolymerization of *n*-butyl acrylate.

3.7.2 Mixed Anionic and Nonionic Surfactant Systems

Chen et al. [90] examined the general validity of the Smith–Ewart theory in the emulsion polymerization of styrene stabilized by a mixed anionic and nonionic system of sodium dodecyl sulfate and NP-40. The CMCs of the mixed surfactant systems are determined for various compositions at 25 °C and 80 °C, and the experimental data are well-described by the regular solution model for mixed micelles. These mixed micelles exhibit a quite nonideal behavior, especially at lower temperature. The effect of the mixed surfactants on particle nucleation is demonstrated by a series of emulsion polymerizations of styrene. Adding only a small amount of sodium dodecyl sulfate into the polymerization system dramatically increases the final number of latex particles per unit volume of water and, therefore, reduces the ultimate latex particle size. Furthermore, the polymerization system stabilized by the mixed surfactants does not follow the conventional Smith–Ewart theory when the level of NP-40 is relatively high. Chern et al. [91] then showed that the polymerization system follows the Smith–Ewart theory only when the level of NP-40 in the mixture of sodium dodecyl sulfate and NP-40 is less than 30 wt%. However, the polymerization system starts to deviate from the Smith–Ewart theory significantly when the level of NP-40 (in the stablizer mixture) is higher than 50 wt%. The steric stabilization effect provided by NP-40 alone is not strong enough to prohibit the interactive latex particles from flocculation. On the other hand, the mixed anionic and nonionic surfactants significantly improve the colloidal stability of the polymerization system via the synergistic effects provided by both the electrostatic and steric stabilization mechanisms and, thus, effectively retard the limited particle flocculation process. The mixed surfactants of sodium dodecyl sulfate/NP-40 = 20/80 (w/w) is probably the best choice because it results in the best reproducibility of the experiments and the fastest rate of polymerization. Chern et al. [92] studied the effect of the initiator (sodium persulfate) concentration on the emulsion polymerization of styrene stabilized by sodium dodecyl sulfate and NP-40. The relationship that the number of latex particles nucleated per unit volume of water is proportional to the surfactant and initiator concentrations to the 0.6 and 0.4 powers, respectively, is only applicable to the polymerization system in the absence of NP-40. At an initiator concentration of 1.38×10^{-3} M, the emulsion polymerizations with 0, 50, and 80 wt% NP-40 result in comparable latex particle sizes and relatively monodisperse particle size distributions throughout the reaction. On the other hand, emulsion polymerizations stabilized by NP-40 alone show the largest latex particle sizes along with the broadest particle size distributions. This is attributed to the long particle nucleation period and/or the limited particle flocculation process. The rate of polymerization increases with increasing initiator concentration for emulsion polymerizations stabilized by sodium dodecyl sulfate. On the other hand, the rate of polymerization remains relatively constant with an increase in the initiator concentration for emulsion polymerizations stabilized by NP-40 alone. As for the emulsion polymerizations stabilized by 50 or 80 wt% NP-40, the rate of polymerization first increases to a maximum and then decreases with increasing initiator concentration.

Unzueta and Forcada [93] developed a mechanistic model for the emulsion copolymerization of methyl methacrylate and *n*-butyl acrylate stabilized by mixed anionic and nonionic surfactants, which was verified by the experimental data. This model is based on the mass and population balances of precursor particles and the moments of particle size distribution. It is sensitive to such parameters as the composition of mixed surfactants and the total surfactant concentration. A competitive particle nucleation mechanism is incorporated into the model to successfully simulate the evolution of particle nuclei during polymerization.

In emulsion polymerization, particle nucleation mechanisms have been the major focus of numerous studies over the past 60 years. These theoretical and experimental results continue to advance our knowledge about the general features of particle nucleation. Conventional polymerization of monomer emulsions involves nucleation and growth of latex particles in a heterogeneous reaction system. Depending on the recipes and reaction conditions, one or more than one of the particle nucleation mechanisms (micellar nucleation, homogeneous nucleation, and coagulative nucleation) can be operative in emulsion polymerization. The pioneering studies on micellar nucleation and homogeneous nucleation attempted to quantitatively describe the formation of particle nuclei in emulsion polymerizations of a variety of monomers. These two particle nucleation mechanisms seem to work quite well in this regard. Nevertheless, at present, the particle nucleation processes are still not well understood, and the issue of the colloidal stability of latex particles makes the situation even more complicated. How to distinguish between the particle nucleation mechanisms and quantitatively determine the fraction of latex particles originating from each nucleation mechanism still remains a great challenge to polymer chemists.

REFERENCES

1. R. G. Gilbert, *Emulsion Polymerization: A Mechanistic Approach*, Academic Press, London, 1995.
2. R. M. Fitch, *Polymer Colloids: A Comprehensive Introduction*, Academic Press, London, 1997.
3. P. A. Lovell and M. S. El-Aasser (Eds.), *Emulsion Polymerization and Emulsion Polymers*, John Wiley & Sons, West Sussex, 1997.
4. H. Y. Erbil, *Vinyl Acetate Emulsion Polymerization and Copolymerization with Acrylic Monomers*, CRC Press LLC, Boca Raton, FL, 2000.
5. C. S. Chern, in *Encyclopedia of Surface and Colloid Science*, A. Hubbard (Ed.), Marcel Dekker, New York, 2002, pp. 4220–4241.
6. A. van Herk, *Chemistry and Technology of Emulsion Polymerization*, Blackwell Publishing, Oxford, 2005.
7. M. Nomura, H. Tobita, and K. Suzuki, Adv. Polym. Sci. **175**, 1 (2005).
8. C. S. Chern, Prog. Polym. Sci. **31**, 443 (2006).
9. W. D. Harkins, J. Chem. Phys. **13**, 381 (1945).
10. W. D. Harkins, J. Chem. Phys. **14**, 47 (1946).
11. W. D. Harkins, J. Am. Chem. Soc. **69**, 1428 (1947).
12. W. V. Smith and R. H. Ewart, J. Chem. Phys. **16**, 592 (1948).
13. W. V. Smith, J. Am. Chem. Soc. **70**, 3695 (1948).
14. W. V. Smith, J. Am. Chem. Soc. **71**, 4077 (1949).
15. J. L. Gardon, J. Polym. Sci. A-1 **6**, 623 (1968).
16. J. L. Gardon, J. Polym. Sci. A-1 **6**, 643 (1968).
17. K. E. Barrett, *Dispersion Polymerization in Organic Media*, Wiley, New York, 1975.
18. J. Ugelstad and F. K. Hansen, Rubber Chem. Technol. **49**, 536 (1976).
19. F. K. Hansen and J. Ugelstad, J. Polym. Sci. Polym. Chem. **16**, 1953 (1978).
20. M. Harada, M. Nomura, H. Kojima, W. Eguchi, and S. Nagata, J. Appl. Polym. Sci. **16**, 811 (1972).
21. M. Nomura, M. Harada, W. Eguchi, and S. Nagata, in *Emulsion Polymerization*, I. Piirma and J. L. Gardon, Eds., ACS Symposium Series, No. 24, Washington, D.C., 1976, p. 102.
22. E. Unzueta and J. Forcada, J. Appl. Polym. Sci. **66**, 445 (1997).
23. P. H. H. Araujo, J. C. de la Cal, J. M. Asua, and J. C. Pinto, Macromol. Theor. Simul. **10**, 769 (2001).
24. C. Sayer, M. Palma, and R. Giudici, Eng. Chem. Res. **41**, 1733 (2002).
25. F. K. Hansen and J. Ugelstad, Makromol. Chem. **180**, 2423 (1979).
26. C. P. Roe, Ind. Eng. Chem. **60**, 20 (1968).

27. W. J. Priest, J. Phys. Chem. **56**, 1977 (1952).
28. R. M. Fitch and C. H. Tsai, in *Polymer Colloids*, R. M. Fitch (Ed.), Plenum Press, New York-London, 1971, p. 73.
29. R. M. Fitch and C. H. Tsai, in *Polymer Colloids*, R. M. Fitch (Ed.), Plenum Press, New York, 1971, p. 103.
30. R. M. Fitch, Br. Polym. J. **5**, 467 (1973).
31. R. M. Fitch, in *Emulsion Polymer and Emulsion Polymerization*, D. R. Bassett and A. E. Hamielec (Eds.), ACS Symposium Series, No. 165, Washington, D.C., 1981, p. 1.
32. F. K. Hansen and J. Ugelstad, in *Emulsion Polymerization*, I. Piirma (Ed.), Academic, New York, 1982, p. 51.
33. G. Lichti, R. G. Gilbert, and D. H. Napper, J. Polym. Sci., Polym. Chem. **21**, 269 (1983).
34. P. J. Feeney, D. H. Napper, and R. G. Gilbert, Macromolecules **17**, 2520 (1984).
35. P. J. Feeney, D. H. Napper, and R. G. Gilbert, J. Colloid Interface Sci. **107**, 159 (1985).
36. P. J. Feeney, E. Geissler, R. G. Gilbert, and D. H. Napper, J. Colloid Interface Sci. **121**, 508 (1988).
37. F. K. Hansen and J. Ugelstad, J. Polym. Sci., Polym. Chem. **17**, 3033 (1979).
38. F. K. Hansen and J. Ugelstad, J. Polym. Sci., Polym. Chem. **17**, 3047 (1979).
39. F. K. Hansen and J. Ugelstad, J. Polym. Sci., Polym. Chem. **17**, 3069 (1979).
40. G. W. Poehlein, in *Emulsion Polymerization*, I. Piirma (Ed.), Academic Press, New York, 1982.
41. Z. Song and G. W. Poehlein, J. Macromol. Sci.-Chem. **25**, 403 (1988).
42. Z. Song and G. W. Poehlein, J. Macromol. Sci.-Chem. **25**, 1587 (1988).
43. M. Arai, K. Arai, and S. Saito, J. Polym. Sci., Polym. Chem. **17**, 3655 (1979).
44. S. P. Chatterjee, M. Banerjee, and R. S. Konar, Indian J. Chem. **14A**, 836 (1976).
45. J. Vanderhoff, J. Polym. Sci. Symp. **72**, 161 (1985).
46. N. Sutterlin, in *Polymer Colloids II*, R. M. Fitch (Ed.), Plenum Press, New York and London, 1980, p. 583.
47. Z. Song and G. W. Poehlein, J. Colloid Interface Sci. **128**, 486 (1989).
48. Z. Song and G. W. Poehlein, J. Colloid Interface Sci. **128**, 501 (1989).
49. J. Brandrup and E. H. Immergut, *Polymer Handbook*, Wiley-Interscience, New York, 1975.
50. I. M. Kolthoff and A. I. Medalia, J. Polym. Sci. **5**, 391 (1950).
51. F. A. Bovey, I. M. Kolthoff, A. I. Medalia, and E. J. Meehan, *Emulsion Polymerization*, Interscience, New York, 1955.
52. B. M. E. Van der Hoff, in Advances in Chemistry Series, No. 34, ACS Publication, Washington, D.C., 1962, p. 6.
53. S. A. Chen and H. S. Chang, J. Polym. Sci. **23**, 2615 (1985).
54. R. L. Zollars, J. Appl. Polym. Sci. **24**, 1353 (1979).
55. A. S. Dunn and L. C.-H. Chong, Br. Polym. J. **2**, 49 (1970).
56. R. M. Fitch, T. H. Palmgren, T. Auyogi, and A. Zuikov, J. Polym. Sci., Symp. **72**, 221 (1985).

57. F. K. Hansen and J. Ugelstad, J. Polym. Sci., Polym. Chem. **17**, 3047 (1979).
58. S. T. Wang and G. W. Poehlein, J. Appl. Polym. Sci. **50**, 2173 (1993).
59. S. T. Wang and G. W. Poehlein, J. Appl. Polym. Sci. **51**, 593 (1994).
60. B. Thomson, Z. Wang, A. Paine, G. Lajoie, and A. Rudin, J. Polym. Sci., Polym. Chem. **33**, 2297 (1995).
61. Y. M. Wang and C. Y. Pan, Colloid Polym. Sci. **277**, 658 (1999).
62. J. L. Ou, J. K. Yang, and H. Chen, Eur. Polym. J. **37**, 789 (2001).
63. C. Yan, S. Cheng, and L. Feng, J. Polym. Sci., Polym. Chem. **37**, 2649 (1999).
64. A. R. Mahdavian and M. Abdollahi, Polymer **45**, 3233 (2004).
65. J. Zhang, Q. Zou, X. Li, and S. Cheng, J. Appl. Polym. Sci. **89**, 2791 (2003).
66. P. K. Sahoo and R. Mohapatra, Eur. Polym. J. **39**, 1839 (2003).
67. K. Tauer, R. Deckwer, I. Kuhn, and C. Schellenberg, Colloid Polym. Sci. **277**, 607 (1999).
68. L. Varela de la Rosa, E. D. Sudol, M. S. El-Aasser, and A. Klein, J. Polym. Sci., Polym. Chem. **34**, 461 (1996).
69. L. Varela de la Rosa, E. D. Sudol, M. S. El-Aasser, and A. Klein, J. Polym. Sci., Polym. Chem. **37**, 4066 (1999).
70. L. Varela de la Rosa, E. D. Sudol, M. S. El-Aasser, and A. Klein, J. Polym. Sci., Polym. Chem. **37**, 4073 (1999).
71. J. Herrera-Ordonez and R. Olayo, J. Polym. Sci., Polym. Chem. Ed. **38**, 2201 (2000).
72. J. Herrera-Ordonez and R. Olayo, J. Polym. Sci., Polym. Chem. Ed. **38**, 2219 (2000).
73. J. Herrera-Ordonez and R. Olayo, J. Polym. Sci., Polym. Chem. Ed. **39**, 2547 (2001).
74. M. Nomura, U. S. Satpathy, Y. Kouno, and K. Fujita, J. Polym. Sci., Polym. Lett. **26**, 385 (1988).
75. M. Nomura, K. Takahashi, and K. Fujita, Macromol. Chem.-M. Symp. **35**/36, 13 (1990).
76. I. Kuhn and K. Tauer, Macromolecules **28**, 8122 (1995).
77. K. Tauer, and R. Deckwer, Acta Polym. **49**, 411 (1998).
78. S. Kozempel, K. Tauer, and G. Rother, Polymer **46**, 1169 (2005).
79. C. S. Chern and C. H. Lin, Polymer **40**, 139 (1998).
80. C. S. Chern and C. H. Lin, Polymer **41**, 4473 (2000).
81. S. Rudschuck, J. Adams, and J. Fuhrmann, Phys. Res. B **151**, 341 (1999).
82. K. Tauer, H. Muller, C. Schellenberg, and L. Rosengarten, Colloid Surface A: Physicochem. Eng. Aspects **153**, 143 (1999).
83. L. Varela De La Rosa, E. D. Sudol, V. L. Dimonie, A. Klein, and M. S. El-Aasser, J. Polym. Sci. Polym. Chem. **37**, 4066 (1999).
84. L. Varela De La Rosa, E. D. Sudol, M. S. El-Aasser, and A. Klein, J. Polym. Sci., Polym. Chem. **37**, 4073 (1999).
85. S. Sajjadi, Macromol. Rapid Commun. **25**, 882 (2004).
86. E. Ozdeer, E. D. Sudol, M. S. El-Aasser, and A. Klein, J. Polym. Sci., Polym. Chem. **35**, 3837 (1997).

87. I. Capek and J. Chudej, Polym. Bull. **43**, 417 (1999).
88. S. Y. Lin, I. Capek, T. J. Hsu, and C. S. Chern, J. Polym., Sci. Polym. Chem. **37**, 4422 (1999).
89. K. Ouzineb, M. Fortuny Heredia, C. Graillat, and T. F. McKenna, J. Polym. Sci., Polym. Chem. **39**, 2832 (2001).
90. L. J. Chen, S. Y. Lin, C. S. Chern, and S. C. Wu, Colloid Surface A: Physicochem. Eng. Aspects **122**, 161 (1997).
91. C. S. Chern, S. Y. Lin, L. J. Chen, and S. C. Wu, Polymer **38**, 1977 (1997).
92. C. S. Chern, S. Y. Lin, S. C. Chang, J. Y. Lin, and Y. F. Lin, Polymer **39**, 2281 (1998).
93. E. Unzueta and J. Forcada, J. Appl. Polym. Sci. **66**, 445 (1997).

4

EMULSION POLYMERIZATION KINETICS

Formation of particle nuclei plays a crucial role during the early stage of emulsion polymerization (Interval I) and this subject was the focus of Chapter 3. Particle nucleation also has a significant impact on the subsequent consumption of monomer within the growing polymer particles. After the particle nucleation process is completed, the number of latex particles (i.e., reaction loci) per unit volume of water remains relatively constant to the end of polymerization. The propagation reaction of free radicals with monomer molecules takes place primarily in these monomer-swollen polymer particles. In general, free radical polymerization of monomer begins with the generation of initiator radicals by the thermal decomposition of the persulfate (or other water-soluble) initiator in the continuous aqueous phase. This is followed by the propagation of these initiator radicals with monomer molecules dissolved in water before entering the monomer-swollen polymer particles to initiate or deactivate the chain addition polymerization therein. Free radicals also can be transported out of the latex particles and reabsorbed by the latex particles during polymerization. Emulsified monomer droplets serve primarily as discrete reservoirs to supply the growing latex particles with monomer and surfactant species due to the very small monomer droplet surface area. The majority of monomer is consumed in this particle growth stage ranging from ~10–20% to 60% monomer conversion. The particle growth stage (Interval II) ends when the thermodynamic driving force for the transport of monomer from the monomer droplets, which now contain some polymer, becomes zero in the polymerization system. The objective of this chapter is to provide the reader with a fundamental understanding of emulsion polymerization kinetics,

Principles and Applications of Emulsion Polymerization, by Chorng-Shyan Chern

especially those events related to transport phenomena of free radicals in the heterogeneous reaction system and particle growth mechanisms. These subjects are crucial to the industrial design of efficient semibatch and continuous emulsion polymerization processes and manufacture of latex products with satisfactory performance properties.

4.1 EMULSION POLYMERIZATION KINETICS

4.1.1 Smith–Ewart Theory

Smith–Ewart Case 2 kinetics [1] has been widely used to calculate the rate of polymerization (R_p):

$$R_p = k_p[\mathrm{M}]_p(\mathbf{n}N_p/N_A) \tag{4.1}$$

where k_p is the propagation rate constant, $[\mathrm{M}]_p$ is the concentration of monomer in the polymer particles, **n** is the average number of free radicals per particle, N_p is the number of latex particles (i.e., reaction loci) per unit volume of water, and N_A is the Avogadro number. Among the kinetic parameters appearing in Eq. (4.1), perhaps, the most difficult ones to predict are the total number of latex particles per unit volume of water (N_p) and the average number of free radicals per particle (**n**). As aforementioned, N_p is primarily determined in the particle nucleation stage (Interval I) and the kinetic models developed for the calculation of **n** are the major focus of this chapter.

The parameter **n** is defined as

$$\begin{aligned}\mathbf{n} &= \sum_{i=0}^{\infty} iN_i \Big/ \sum_{i=0}^{\infty} N_i \\ &= \sum_{i=0}^{\infty} iN_i/N_p \end{aligned} \tag{4.2}$$

where N_i is the number of latex particles containing i free radicals ($i = 0, 1, 2, \ldots, \infty$) and N_p is the total number of latex particles per unit volume of water. The value of N_i can be calculated by the following population balance equation [1, 2]:

$$\begin{aligned} dN_i/dt = &(\rho_a/N_p)N_{i-1} + (k_{\mathrm{des}}a_p/v_p)(i+1)N_{i+1} + (k_{t,p}/v_p)[(i+2)(i+1)]N_{i+2} \\ &-\{\rho_a/N_p + (k_{\mathrm{des}}a_p/v_p)i + k_{t,p}/v_p[i(i-1)]\}N_i \end{aligned} \tag{4.3}$$

where ρ_a is the overall rate of absorption of free radicals by the latex particles, k_{des} is the rate constant for desorption of free radicals out of the latex particles, $k_{t,p}$ is the rate constant for bimolecular termination of two adjacent free radicals inside the latex particle, and a_p and v_p are the surface area and volume of a single latex particle, respectively. The parameter ρ_a is defined as

$$\rho_a = \rho_i + k_{\mathrm{des}}\mathbf{n}N_p - 2k_{t,w}[R^*]_w^2 \tag{4.4}$$

$$\rho_i = 2fk_d[\mathrm{I}] \tag{4.5}$$

where ρ_i is the rate of generation of free radicals in the continuous aqueous phase, f is the initiator efficiency factor ($0 \le f \le 1$), k_d is the initiator decomposition rate constant, [I] is the concentration of initiator in the continuous aqueous phase, $k_{t,w}$ is the rate constant for bimolecular termination of two adjacent free radicals in the continuous aqueous phase, and $[\mathrm{R}^*]_w$ is the total concentration of free radicals in water. Equation (4.4) indicates that desorption of free radicals out of the latex particles enhances the rate of absorption of free radicals by the latex particles, whereas the consumption of free radicals by the aqueous termination reactions reduces the frequency of entry of free radicals into the latex particles.

This population balance approach indicates that transport of free radicals in the heterogeneous emulsion polymerization system plays a crucial role in the reaction kinetics. For example, the rate of change in the number of latex particles containing one free radical per unit volume of water (N_1) is equal to the summation of the rate of absorption of one free radical by the latex particles containing zero free radical, the rate of desorption of one free radical out of the latex particles containing two free radicals, and the rate of termination of two free radicals within the latex particles containing three free radicals minus the summation of the rate of absorption of one free radical by the latex particles containing one free radical, the rate of desorption of one free radical out of the latex particles containing one free radical, and the rate of termination of two free radicals within the latex particles containing one free radical. It should be noted that in this case the rate of termination of two free radicals within the latex particles containing one free radical is equal to zero. This is simply because at least two free radicals are required for the bimolecular termination reactions to take place within a latex particle. As another example, the rate of changes in the number of latex particles containing zero free radical per unit volume of water (N_0) is equal to the summation of the rate of desorption of one free radical out of the latex particles containing one free radical and the rate of termination of two free radicals within the latex particles containing two free radicals minus the rate of absorption of one free radical by the latex particles containing zero free radical. Thus, the average number of free radicals per particle (**n**) is primarily determined by absorption of free radicals by the latex particles, desorption of free radicals out of the particles, and bimolecular termination of free radicals inside the particles (Figure 4.1). With the knowledge of the kinetic parameters k_p, f, k_d, k_{des}, $k_{t,p}$, and $k_{t,w}$, the number of latex particles per unit volume of water (N_p), the volume of a latex

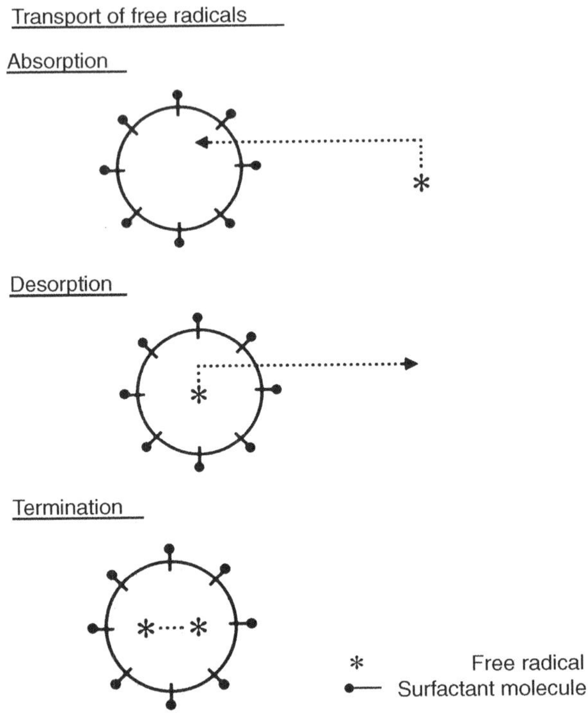

Figure 4.1. Schematic representation of the transport of free radicals between the continuous aquesous phase and the polymer particle phase.

particle (v_p), the concentration of monomer in the latex particles ($[M]_p$), the concentration of initiator in water ([I]), and the concentration of total free radicals in water ($[R^*]_w$), Eqs. (4.2)–(4.5) can be solved simultaneously to calculate **n** and then Eq. (4.1) can be used to predict the rate of polymerization (R_p).

A well-known emulsion polymerization kinetic model can be developed based on the following assumptions:

(a) Nucleation and coagulation of latex particles do not occur and the number of latex particles per unit volume of water remains constant during emulsion polymerization.
(b) The latex particle size distribution is relatively monodisperse.
(c) Desorption of free radicals out of the latex particles does not take place.
(d) Bimolecular termination of a polymeric radical inside the latex particle upon the entry of an oligomeric radical from the continuous aqueous phase is instantaneous.

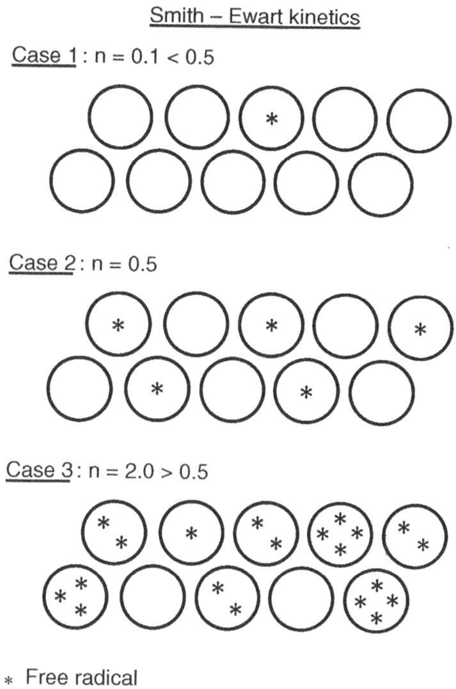

Figure 4.2. Schematic representation of the Smith–Ewart kinetics Cases 1–3.

These assumptions then lead to a scenario that, at any moment, the monomer-swollen polymer particles contain either only one free radical (active) or none (idle). Under these circumstances, a value of **n** equal to 0.5 is achieved for the polymerization systems that follow the Smith–Ewart Case 2 kinetics. In addition, the concentration of monomer in the polymer particles does not vary to any extent with the progress of polymerization in the presence of monomer droplets. As a result, a steady polymerization rate is attained during Interval II. Furthermore, the polymerization kinetics is strictly controlled by the population of polymer particles available for consuming monomer. Smith–Ewart Case 2 kinetics has been successfully applied to emulsion polymerizations of relatively water-insoluble monomers such as styrene and butadiene.

At pesudo-steady state (i.e., $dN_i/dt = 0$), three limiting cases can be obtained from this system of first-order ordinary differential equations (Figure 4.2).

Case 1: $\rho_a/N_p << k_{des}\, a_p/v_p$ (fast desorption)

$$\mathbf{n} = \rho_a v_p/(k_{des} a_p N_p) << 0.5 \tag{4.6}$$

Case 2: $k_{des}\, a_p/v_p << \rho_a/N_p << k_{t,p}/v_p$ (no desorption and fast termination)

$$\mathbf{n} = 0.5 \tag{4.7}$$

Case 3: $\rho_a/N_p >> k_{t,p}/v_p$ (fast absorption and slow termination)

$$\mathbf{n} = [\rho_a v_p/(2k_{t,p}N_p)^{1/2}] >> 1 \tag{4.8}$$

It should be noted that the pseudo-steady-state assumption adopted is quite reasonable because the concentration of free radicals in the reaction system is very low and the reactivity of free radicals is extremely high.

4.1.2 Pioneering Kinetic Models for Predicting Average Number of Free Radicals per Particle

Based on the Smith–Ewart theory [1], Stockmayer [3] derived the following equations to calculate the steady value of **n** when desorption of free radicals out of the latex particles is insignificant (i.e., $k_{des} = 0$).

$$\mathbf{n} = a/4[I_0(a)/I_1(a)] \tag{4.9}$$

$$a = (8\alpha)^{1/2} \tag{4.10}$$

$$\alpha = \rho_a v_p/(k_{t,p}N_p) \tag{4.11}$$

where $I_0(a)$ and $I_1(a)$ are the Bessel functions of the first kind of order 0 and 1, respectively. O'Toole [4] extended this approach to take into account desorption of free radicals out of the particles and obtained the following equations:

$$\mathbf{n} = a/4[I_m(a)/I_{m-1}(a)] \tag{4.12}$$

$$m = k_{des}a_p/k_{tp} \tag{4.13}$$

A schematic representation of the log(**n**) versus log(α) profile is shown in Figure 4.3.

At low values of α, the emulsion polymerization system is characterized by (a) the very slow overall rate of absorption of free radicals by the latex particles and/or the very large population of latex particles (i.e., ρ_a/N_p <<) and (b) the very large rate constant for desorption of free radicals out of the latex particles and/or the very large ratio of the surface area to volume of a single latex particle (i.e., $k_{des}a_p/v_p$ >>). In this regime, desorption of free radicals out of the latex particles plays an important role in the emulsion polymerization kinetics. The value of **n** is smaller than 0.5 [Smith–Ewart Case 1 kinetics, Eq. (4.6)] and **n** increases rapidly with increasing α. At medium values of α, the emulsion polymerization system is characterized by (a) the very small rate

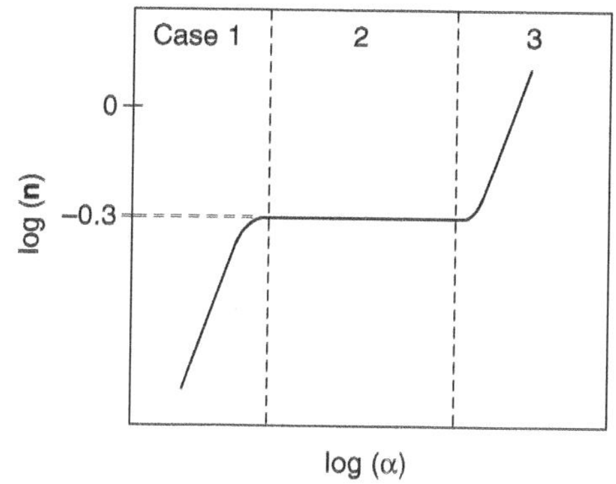

Figure 4.3. Schematic representation of the log (**n**) versus log (α) profile for a typical emulsion polymerization system. The three limiting cases of the Smith–Ewart kinetic model are also indicated in this plot.

constant for desorption of free radicals out of the latex particles and/or the medium ratio of the surface area to volume of a single latex particle (i.e., k_{des} a_p/v_p <<) and (b) the comparable ratio of the overall rate of absorption of free radicals by the latex particles to the number of latex particles per unit volume of water (i.e., medium values of ρ_a/N_p) and (c) the very large rate constant for bimolecular termination of two adjacent free radicals inside the latex particle and/or the relatively small volume of a single latex particle (i.e., $k_{t,p}/v_p$ >>). In this plateau region, desorption of free radicals out of the latex particles is insignificant and the bimolecular termination of free radicals within the latex particles is instantaneous. As a result, the latex particles contain either only one free radical (active) or none (idle) and **n** is equal to 0.5 [Smith–Ewart Case 2 kinetics, Eq. (4.7)]. At very large values of α, the emulsion polymerization system is then characterized by (a) the very large ratio of the overall rate of absorption of free radicals by the latex particles to the number of latex particles per unit volume of water (i.e., ρ_a/N_p >>) and (b) the very small rate constant for bimolecular termination of two adjacent free radicals inside the latex particle due to the gel effect and/or the relatively large volume of a single latex particle (i.e., $k_{t,p}/v_p$ <<). In this regime, **n** increases significantly with increasing α. Under such circumstances, **n** is much larger than one [Smith–Ewart Case 3 kinetics, Eq. (4.8)]. Typical examples are emulsion polymerizations of methyl methacrylate or styrene carried out at temperatures lower than their glass transition temperatures [5–7].

Ugelstad et al. [8, 9] then incorporated the kinetic events of the bimolecular termination reaction taking place in the continuous aqueous phase and reabsorption of the desorbed free radicals by the latex particles into the O'Toole model, and the mass balance equation for free radicals in the continuous aqueous phase can be expressed as

$$\alpha = \alpha' + m\mathbf{n} - Y\alpha^2 \tag{4.14}$$

$$\alpha' = \rho_i v_p/(k_{t,p} N_p) \tag{4.15}$$

$$Y = 2N_A k_{t,w} k_{t,p}/(k_c^2 v_p N_p) \tag{4.16}$$

where k_c is the rate coefficient for the capture of free radicals by the latex particles. The authors then solved the above simultaneous equations for the average number of free radicals per particle and plotted the values of **n** versus α' at various values of Y and m.

The basic framework of emulsion polymerization mechanisms and kinetics is primarily built on the aforementioned pioneering studies, and many other excellent contributions appeared thereafter. Several very useful empirical or approximate equations for calculating **n** were also developed for the emulsion polymerization system in the absence of the bimolecular termination of free radicals in the continuous aqueous phase (i.e., $Y = 0$). For example:

(a) When m is equal to zero, then the following equation can be used to calculate **n** [9]:

$$\mathbf{n} = (1/4 + \alpha'/2)^{1/2} \tag{4.17}$$

(b) When the bimolecular termination of free radicals inside the latex particles is instantaneous and the value of **n** is smaller than or equal to 0.5, then **n** can be estimated by the following equation [10–12]:

$$\mathbf{n} = 1/2\{-\alpha'/m + [(\alpha'/m)^2 + 2\alpha'/m]^{1/2}\} \tag{4.18}$$

The transient behavior of **n** was investigated by a number of studies (e.g., see references 13–16). However, the pseudo-steady-state assumption is generally accepted for the mole balance of free radicals in the polymerization system due to the very high reactivity and low concentration of free radicals. This approach can simplify the analysis of the emulsion polymerization kinetics and save computation time significantly.

It should be noted that the rate of absorption of free radicals by the latex particles from the continuous aqueous phase (ρ_a or α') is not equal to the rate of generation of free radicals in the continuous aqueous phase (ρ_i or α) when desorption of free radicals out of the latex particles (m) and/or the bimolecular termination of free radicals in the continuous aqueous phase (Y) cannot be neglected in the emulsion polymerization system. In addition to the particle nucleation mechanisms discussed in Chapter 3, to gain a fundamental understanding of transport of free radicals in the heterogeneous reaction system (e.g., absorption of free radicals by the latex particles, desorption of free radicals out of the latex particles and reabsorption of the desorbed free radicals by the latex particles) is thus required to predict the emulsion polymerization kinetics [e.g., Eqs. (4.1), (4.12), and (4.14)]. It is the subject that will be considered in detail in the following subsections.

4.2 ABSORPTION OF FREE RADICALS BY LATEX PARTICLES

A water-soluble initiator is generally used in initiating the free radical polymerization of monomer emulsions. The primary reaction loci are the monomer-swollen polymer particles, which are nucleated during the early stage of polymerization. Initiator radicals (such as $-SO_4^{-*}$) are continuously generated in the continuous aqueous phase throughout the emulsion polymerization. The initiator radicals become surface-active immediately after the propagation reaction of these highly reactive species with several monomer molecules dissolved in the continuous aqueous phase. Apparently, the relatively hydrophobic oligomeric radicals must enter the monomer-swollen polymer particles in order to initiate the major free radical polymerization therein. The primary driving forces responsible for this transport phenomenon are (a) the free radical concentration gradient established between the continuous aqueous phase and the discrete polymer particle phase and (b) the increased surface activity of oligomeric radicals with their chain length. Thus, the relatively hydrophobic oligomeric radicals can diffuse toward the vicinity of the latex particle surfaces, adsorb onto the particle surfaces and/or propagate with monomer molecules within the particle surface layer. They can also penetrate deep into the interior of the latex particles and then participate in the major free radical polymerization reactions therein. It is noteworthy that the probability for oligomeric radicals to be absorbed by the emulsified monomer droplets is rather slim because the total monomer droplet surface area is negligible compared to that of monomer-swollen polymer particles.

A number of models dealing with absorption of free radicals by the latex particles were proposed. They are (a) the collision-controlled model [1, 17, 18], (b) the diffusion-controlled model [19], (c) the surfactant displacement model [20], (d) the colloidal model [21], and (e) the propagation-controlled model [22, 23]. The dependence of the rate constant for absorption of free radicals by the latex particles on the particle diameter (d_p) predicted by these models is summarized in Table 4.1 [24]. At present, the most widely accepted models

Table 4.1. Dependence of the Rate Constant for Absorption of Free Radicals by the Latex Particles Predicted by the Models [24]

Model	Dependence on d_p	References
Collision-controlled	None	1, 17, 18
Diffusion-controlled	d_p^2	19
Surfactant displacement	d_p	20
Colloidal	d_p	21
Propagation-controlled	None	22, 23

are the collision-controlled model, the diffusion-controlled model, and the propagation-controlled model.

4.2.1 Collision- and Diffusion-Controlled Models

Smith and Ewart [1] proposed that the rate of absorption of free radicals by a latex particle is expressed as follows:

$$\begin{aligned} \rho_a/N_p &= 2\pi D_w d_p[R^*]_w \\ &= k_{e,p}[R^*]_w \end{aligned} \tag{4.19}$$

where D_w is the diffusion coefficient of oligomeric radicals in the continuous aqueous phase and $k_{e,p}$ is the rate constant for the absorption of free radicals by the latex particles. Equation (4.19) indicates that transport of free radicals from the continuous aqueous phase to the interior of the latex particles is achieved simply by the molecular diffusion of an oligomeric radical from the continuous aqueous phase, across the water–particle interface, into the interior of the latex particle containing no free radicals. It should be noted that the rate constant for absorption of free radicals by the latex particles was assumed to be proportional to the square of the latex particle diameter (i.e., proportional to the latex particle surface area) in the work of Smith and Ewart [1]. In other words, the probability for the collision between the latex particles and free radicals in the continuous aqueous phase and the subsequent absorption of these free radicals by the latex particles to occur increases linearly with increasing total particle surface area. This collision-controlled model was then adopted by Gardon [17, 18] to predict the rates of absorption of free radicals by the monomer-swollen micelles and particle nuclei, respectively, in the particle nucleation stage.

Harada et al. [25] developed a mechanistic model for predicting the styrene emulsion polymerization kinetics. Equation (4.19) was adopted to calculate the rate of absorption of free radicals by the latex particles or the monomer-swollen micelles. Comparing the model predictions with the experimental data, an approximate value of 10^3 for the ratio $k_{e,p}/k_{e,m}$ was obtained, where $k_{e,m}$ is the rate constant for the absorption of free radicals by the monomer-swollen micelles. Nevertheless, this value of $k_{e,p}/k_{e,m}$ is at least two orders of magnitude greater than that predicted by Eq. (4.19). This is because the ratio $k_{e,p}/k_{e,m}$ is equal to d_p/d_m according to Eq. (4.19), where d_m is the diameter of monomer-swollen micelles. As a first approximation, the value of d_p/d_m is in the order of 10 during the early stage of emulsion polymerization. This implies that the efficiency of absorption of free radicals by the latex particles is about 100 times as large as that of absorption of free radicals by the monomer-swollen micelles. This can be attributed to the postulations that (a) the free energy barrier for free radicals to enter the monomer-swollen micelles is higher compared to the latex particle counterpart and (b) free radicals in the micelles can desorb back into the continuous aqueous phase without initiating the free radical polymerization therein because the volume of a micelle is so small that the mean residence time for a free radical inside the micelle is quite short and the probability for this free radical to propagate with monomer molecules is thus reduced significantly.

The concept of free radical capture efficiency was incorporated into the work of Hansen and Ugelstad [26, 27]. Based on the mechanism of mass transfer with simultaneous chemical reactions, the net rate of absorption of free radicals by a single latex particle (ρ_a/N_p) can be written as

$$\begin{aligned}\rho_a/N_p &= 2\pi D_w d_p[R^*]_w F \\ &= k_{e,p}[R^*]_w\end{aligned} \tag{4.20}$$

where F is the free radical capture efficiency factor defined as follows:

$$1/F = (D_w/\lambda D_p)(X \coth X - 1)^{-1} + W' \tag{4.21}$$

where λ is the equilibrium partition coefficient for free radicals between the continuous aqueous phase and the latex particle phase, D_p is the diffusion coefficient for free radicals in the latex particles, X is defined as $(d_p/2)\{(k_p[\mathrm{M}]_p + k_{t,p}n/v_p)/D_p\}^{1/2}$, n is the number of free radicals in a latex particle, and W' is the potential energy barrier against absorption of free radicals by the latex particles. Equation (4.21) predicts that the value of F for the latex particles containing free radicals ($n \geq 1$) is larger than that for the latex particles containing no free radicals ($n = 0$). Furthermore, F increases with increasing particle size ($F \sim d_p^2$) for the latex particles containing zero free radicals, whereas F first decreases to a minimum and then increases with increasing particle size for the latex particles containing free radicals. Considering the competitive absorption of free radicals by the particle nuclei and micelles containing zero free radicals again, the ratio $k_{e,p}/k_{e,m}$ is equal to $(d_p/d_m)^3 = (10/1)^3 = 10^3 (k_{e,p} \sim d_pF \sim d_p^3)$ and $k_{e,m} \sim d_mF \sim d_m^3$ according to Eq. (4.20). This theoretical estimation is consistent with the experimental work of Harada et al. dealing with the emulsion polymerization of styrene [25].

Another useful expression for the free radical capture efficiency factor (F) is shown below:

$$F = \{k_p[\mathrm{M}]_p + k_{t,p}(n/v_p)\}/\{k_{\mathrm{des}} + k_p[\mathrm{M}]_p + k_{t,p}(n/v_p)\} \tag{4.22}$$

The parameter k_{des} in this equation can be calculated by the following expression [11, 28, 29]:

$$\begin{aligned}k_{\mathrm{des}} &= [2\pi D_w d_p/(m_d v_p)][1 + (\varphi D_w)/(m_d D_p)]^{-1} \\ &= 12 D_w \delta/(m_d d_p^2)\end{aligned} \tag{4.23}$$

where m_d ($= [\mathrm{M}]_p/[\mathrm{M}]_w$) is the partition coefficient for monomer between the latex particle phase and the continuous aqueous phase, $[\mathrm{M}]_w$ is the

Table 4.2. Some Representative Data Regarding the Absorption of Free Radicals by the Monomer-Swollen Micelles and Polymer Particles

System	d_p/d_m	F_m	F_p	$k_{e,p}/k_{e,m}$	Reference
St	10^1			10^3	25
MMA/BA		10^{-5}	10^{-4}		31
VAc		1.0×10^{-5}	3.3×10^{-3}	333	32
VAc				10^4	33
VAc/Veova 10	10^1	1.5×10^{-4}	1.5×10^{-3}	10^2	34

St, styrene; MMA, methyl methacrylate; BA, *n*-butyl acrylate; VAc, vinyl acetate; Veova 10, a commercially available comonomer.

concentration of monomer in water, φ is a constant (in the range of 1–6) that is dependent on the form of $k_{e,p}$ used (e.g., $\varphi = 1$ when Eq. (4.19) is used to describe the mass transfer process inside the latex particles [11, 28–30]), the term $\varphi D_w/(m_d D_p)$ is the ratio of the diffusion resistance on the latex particle side to the diffusion resistance on the water side, and δ ($= [1 + (\varphi D_w)/(m_d D_p)]^{-1}$) is the ratio of the diffusion resistance on the water side to the overall diffusion resistance. Under the condition that the terms $k_p[\mathrm{M}]_p$ and $k_{t,p}(\mathbf{n}/v_p)$ in Eq. (4.22) are negligible compared to k_{des}, this equation then becomes

$$\begin{aligned} F &= k_p[\mathrm{M}]_p / k_{\mathrm{des}} \\ &= \{k_p[\mathrm{M}]_p m_d \delta / (12 D_w)\} d_p^2 \end{aligned} \tag{4.24}$$

The relationship $F \sim d_p^2$ obtained from Eq. (4.24) agrees with the work of Hansen and Ugelstad [26], and thus the ratio $k_{e,p}/k_{e,m} = (d_p/d_{\mathrm{m}})^3 = 10^3$ obtained experimentally from the emulsion polymerization of styrene is further confirmed [25].

Unzueta and Forcada [31] studied the emulsion copolymerization of methyl methacrylate and *n*-butyl acrylate. It was assumed that both micellar nucleation and homogeneous nucleation are operative in this emulsion polymerization system. Based on the experimental data and computer simulation results, the values of the free radical capture efficiency factors for monomer-swollen micelles (F_m) and polymer particles (F_p) that serve as adjustable parameters in the kinetic modeling work are approximately 10^{-5} and 10^{-4}, respectively. The reason for such a difference in the free radical capture efficiency factors is not available yet. Table 4.2 summarizes some representative data regarding the absorption of free radicals by the monomer-swollen micelles and polymer particles obtained from the literature.

4.2.2 Propagation-Controlled Model

Development of the propagation-controlled model for the absorption of the initiator-derived free radicals is based on the following major assumptions [22]:

(a) A free radical in the continuous aqueous phase enters a monomer-swollen polymer particle only when it reacts with a critical number (z) of monomer molecules.

(b) The free radical comprising z monomeric units can successfully enter the latex particles before being terminated with the other free radical in the continuous aqueous phase due to its very fast entry rate.

Thus, generation of free radicals with z monomeric units via the propagation reaction of free radicals with $(z - 1)$ monomeric units with monomer molecules in the continuous aqueous phase is the rate-limiting step for the absorption of free radicals by the latex particles. The rate of absorption of free radicals by a single latex particle ($\rho = \rho_a/N_p$) can be written as follows:

$$\rho = k_{p,w}[IM_{z-1}^{*}]_w[M]_w \tag{4.25}$$

where $k_{p,w}$, $[IM_{z-1}*]_w$, and $[M]_w$ are the propagation rate constant, concentration of the initiator-derived free radicals comprising $(z - 1)$ monomeric units and concentration of monomer in water, respectively. Assuming that the pseudo-steady state is applicable to $[IM_{z-1}*]_w$, Eq. (4.25) can be transformed into the following relationship:

$$\begin{aligned}\rho &= (2k_d[I]/N_p)\{(k_d[I]k_{t,w})^{1/2}/(k_{p,w}[M]_w)+1\}^{1-z} \\ &= (2k_d[I]/N_p)f_e\end{aligned} \tag{4.26}$$

where f_e $(= \{(k_d[I]k_{t,w})^{1/2}/(k_{p,w}[M]_w) + 1\}^{1-z})$ is the initiator efficiency and representative values of z are listed in Table 4.3. For those who are interested in the methodology for experimentally determining ρ, z, and f_e, refer to references 14, and 36–39. For example, the values of f_e for the emulsion polymerization of methyl methacrylate are 0.36 for the anionic initiator potassium

Table 4.3. Representative Values of *z* for the Persulfate Initiator-Derived Free Radicals

	z Value	
Monomer	Maxwell et al., 50 °C [22]	Dong and Sundberg, 25 °C [35]
2-Ethylhexyl acrylate		1
Styrene	2–3	2
Butyl methacrylate	3	2
n-Butyl acrylate	2–3	2
Butadiene	3	2
Ethyl acrylate		4
Methyl methacrylate	4–5	4
Vinyl acetate		5
Methyl acrylate		8
Acrylonitrile		>10 (~12)

persulfate and 0.33 for the cationic initiator 2,2′-azobis(2-amidinopropane) dihydrochloride (V-50) at 70 °C, respectively [37]. These experimental results suggest that only a small fraction (~0.35) of the initiator radicals generated in the continuous aqueous phase can be ultimately captured by the latex particles and then participate the major propagation reaction with monomer molecules therein.

4.2.3 Some Controversial Issues

As mentioned above, the two most popular reaction mechanisms involved in the absorption of free radicals by the monomer-swollen micelles and polymer particles are the diffusion- and propagation-controlled models. Nevertheless, Liotta et al. [39] were inclined to support the collision-controlled model. A dynamic competitive particle growth model was developed to study the emulsion polymerization of styrene in the presence of two distinct populations of latex particles (i.e., bimodal particle size distribution). Comparing the on-line density and particle size data with model predictions suggests that absorption of free radicals by the latex particles follows the collision-controlled mechanism.

One important question about emulsion polymerization mechanisms is which type of free radicals can be captured by the monomer-swollen micelles and polymer particles. It is generally accepted that only a surface-active oligomeric radical terminated with a sulfate end-group can enter a latex particle in the emulsion polymerization of styrene initiated by a persulfate initiator. Tauer and Deckwer [40] used the MALDI-TOF-MS technique to investigate the surfactant-free emulsion polymerization of styrene initiated by a persulfate initiator. In addition to the sulfate end-group, a variety of end-groups can be found in the polymer chains. These experimental results provide supporting evidence of the conclusion that surface activity is not a prerequisite for free radicals (either primary radicals containing –H, –OH, or $-SO_4^-$ end-groups or oligomeric radicals) to be captured by the latex particles.

The role of the particle surface charges or the hydrophilic surface layer of surfactant in the absorption of free radicals by the latex particles is also of great interest to those who are involved in emulsion polymerization. In summary, for emulsion polymerization systems stabilized by conventional surfactants such as sodium dodecyl sulfate, both the particle surface charge density and the ionic strength eventually show an insignificant influence on the entry of free radicals from the continuous aqueous phase into the latex particle phase [21, 41]. Colombie et al. [38] studied the effect of *i*-octylphenoxypolyethoxy ethanol with an average of 40 ethylene oxide units per molecule (Triton X-405) on the rate of absorption of free radicals by the latex particles by varying the ratio of Triton X-405 to sodium dodecyl sulfate in the seeded emulsion polymerization of styrene initiated by potassium persulfate. It was shown that the adsorbed layer of Triton X-405 on the latex particle surface does not serve as a steric barrier toward the incoming free radicals. On the other hand,

Kusters et al. [42] showed that the desorption rate constant for monomeric radicals decreases significantly, compared to anionic surfactant, in the emulsion polymerization stabilized by *n*-nonylphenoxypolyethoxy ethanol with an average of 30 ethylene oxide units per molecule. This result implies that there exists a steric barrier surrounding the latex particle that retards the entry (or exit?) of free radicals for the emulsion polymerization system stabilized by nonionic surfactant. More research efforts are required to reconcile this controversial issue.

The rate coefficient for absorption of free radicals by the latex particles is several orders of magnitude lower than the diffusion-controlled rate coefficient [43]. This can be attributed to the presence of a free energy barrier for the penetration of free radicals into the latex particles and the bimolecular termination of free radicals in the particles. The bimolecular termination of free radicals in the continuous aqueous phase is generally considered insignificant (i.e., $Y = 0$). However, Gilbert and Napper [44] and Lichti et al. [45] illustrated that the aqueous phase termination reaction can have a significant influence on the emulsion polymerization kinetics, especially for the reaction system with a fast initiation rate, a small population of latex particles, and large latex particles. Lee and Poehlein [46] pointed out that the aqueous phase termination reaction does not play an important role in the polymerization system with the kinetic parameter m smaller than one and α' less than ~0.01. A large number of emulsion polymerization systems satisfy such a criterion.

4.3 DESORPTION OF FREE RADICALS OUT OF LATEX PARTICLES

It is generally accepted that desorption of free radicals out of the latex particles is caused by the chain transfer of a polymeric radical to monomer (or a chain transfer agent) in the monomer-swollen polymer particles. As a result, a rather mobile monomeric (or CTA) radical is generated, and this is followed by molecular diffusion of this free radical from the interior of the latex particle, across the particle–water interface, and then into the continuous aqueous phase. Apparently, the desorbed radicals exhibit very different characteristics from those of oligomeric radicals originating from the persulfate initiator radicals. On the other hand, the monomeric radical may participate in the free radical polymerization before being transferred into the continuous aqueous phase. It should be noted that the desorbed monomeric radical might also have the chance to be absorbed again by another latex particle and reinitiate the propagation reaction with monomer molecules therein. Furthermore, the bimolecular termination reaction between two approaching free radicals in the continuous aqueous phase may also happen in some emulsion polymerization systems. As a result of desorption of free radicals out of the latex particles, the average number of free radicals per particle (**n**) decreases, as illustrated by the Smith–Ewart Case 1 kinetics in Figure 4.3. The rate of polymerization (R_p) decreases accordingly.

4.3.1 Desorption of Free Radicals in Emulsion Homopolymerization Systems

Based on the above-mentioned free radical desorption mechanism, the research groups of Ugelstad [8–10] and Nomura [11, 33, 47] independently derived the following relationship for the desorption rate constant (k'_{des}):

$$k'_{\text{des}} = \{(k_{\text{tr},m}/k'_p)[12(\pi/6)^{2/3} D_w/(a + D_w/D_p)]\} + \{(k_{\text{tr},t}[\text{T}]_p/k_{ti}[\text{M}]_p)[12(\pi/6)^{2/3} D_w/(a_t + D_{w,t}/D_{p,t})]\} \tag{4.27}$$

where k'_{des} is defined as $k_{\text{des}}\ (v_p)^{2/3}$ and it is independent of the latex particle size. The kinetic parameters $k_{\text{tr},m}$ and $k_{\text{tr},t}$ are the rate coefficients for the chain transfer of a polymeric radical to monomer and chain transfer agent (e.g., *n*-butyl mercaptan and carbon tetrachloride) inside the particle, respectively, k'_p and k_{ti} are the reinitiation rate coefficients for monomeric radicals and chain transfer agent radicals, respectively, a and a_t are the partition coefficients for monomeric radicals and chain transfer agent radicals between the particle phase and the continuous aqueous phase, respectively, D_p is the diffusion coefficient for monomeric radicals in the particle phase, $[\text{T}]_p$ is the concentration of chain transfer agent in the particles, and $D_{w,t}$ and $D_{p,t}$ are the diffusion coefficients of chain transfer agent radicals in the water and particle phases, respectively. The major assumptions used in the model development are summarized as follows:

(a) Latex particles contain either zero or one free radical.
(b) Bimolecular termination reaction takes place immediately after the absorption of a free radical by the latex particle already containing one free radical.
(c) For the desorbed free radicals, propagation, bimolecular termination, and chain transfer reactions in the continuous aqueous phase are negligible. This implies that all the desorbed free radicals would be absorbed by the latex particles. These monomeric radicals or chain transfer agent radicals would lose their reactivity only after the propagation or bimolecular termination reactions occur inside the latex particles.

A number of studies endeavored to experimentally determine the values of the desorption rate constant. It is also interesting to note that Lee and Poehlein [46, 48, 49] modified the approach of Ugelstad et al. [8, 9] and applied it to emulsion polymerization carried out in a single continuous stirred tank reactor (CSTR) system. The resultant latex particle size distribution data were then used to determine the value of k'_{des}. The k'_{des} data obtained from other literature are summarized in Table 4.4. Significant variations in the values of k'_{des} for the emulsion polymerizations of styrene at 60 °C are observed.

Table 4.4. Values of Desorption Rate Constant Obtained from the Literature

Monomer	Temperature (°C)	k'_{des} (cm^2s^{-1})	References
Styrene	60	6.0×10^{-12}	Theory
Styrene	60	3.9×10^{-14}	50, 51
Styrene	50	$(4.3 \pm 2.9) \times 10^{-13}$	30, 52
Styrene	60	4.8×10^{-13}	44
Styrene	60	$(5.3 \pm 1.8) \times 10^{-13}$	46
Methyl acrylate	50	4×10^{-12}	53
Styrene	70	7.6×10^{-13}	54
Styrene	80	1.1×10^{-12}	54

Reabsorption of the desorbed free radicals by the latex particles may contribute to the emulsion polymerization kinetics, as proposed by Ugelstad and Hansen [19]. According to Gilbert [55], the overall rate of absorption of free radicals by the latex particles (ρ_a) can be written as

$$\rho_a = \rho + p k_{des} \mathbf{n} \tag{4.28}$$

where ρ is the rate of absorption of free radicals by the latex particles in the absence of the desorption event, and p is a parameter that represents the degree of reabsorption of the desorbed free radicals ($-1 \leq p \leq 1$). All of the desorbed free radicals can successfully enter the latex particles provided that p is equal to one. On the other hand, p is equal to −1 when all of the desorbed free radicals terminate with the oligomeric radicals generated in the continuous aqueous phase. It should be noted that ρ is a function of [I] and N_p only.

Asua et al. [56–58] also derived a mechanistic model for predicting the desorption rate coefficient, which considers several possible chemical reactions for the desorbed free radicals in the continuous aqueous phase and the competitive free radical desorption and bimolecular termination reaction in the latex particles containing more than one free radical. Some other representative publications related to the free radical desorption process can be found in references 59–61.

Desorption of free radicals from the latex particles reduces the value of **n** and, therefore, slows the polymer reaction. The influence of the degree of free radical desorption (i.e., $m = k_{des}a_p/k_{tp}$) on the average number of free radicals per particle (**n**) is schematically shown in Figure 4.4. At constant α, **n** decreases significantly with increasing m, which can be attributed to the increase of k_{des}, the increase of a_p (or the decrease of d_p), or the decrease of k_{tp}. This transport phenomenon provides a reasonable explanation for the emulsion polymerization systems exhibiting significant chain transfer reactions [10, 11, 45, 47, 62].

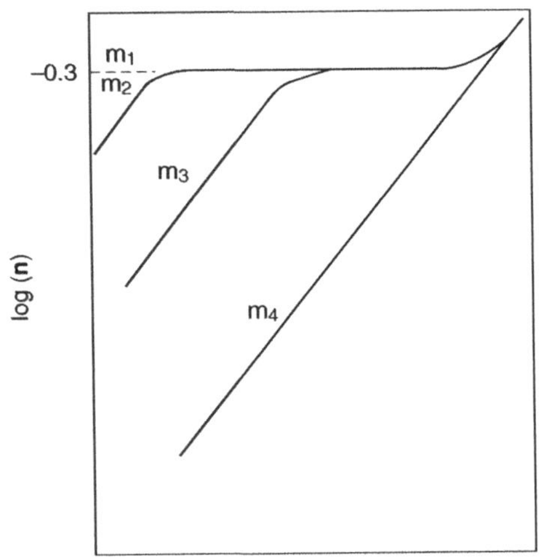

Figure 4.4. Schematic representation of the log (**n**) versus log (α) profiles with different values of m ($m_1 = 0$ (Smith–Ewart kinetics Case 2) $< m_2 < m_3 < m_4$) for a typical emulsion polymerization system.

4.3.2 Desorption of Free Radicals in Emulsion Copolymerization Systems

The pseudo-homopolymerization approach [30, 52] can be used to calculate the average rate coefficient ($\mathbf{k}_{\text{des}}$) for desorption of free radicals out of the latex particles in emulsion copolymerization systems. For a binary polymerization system comprising comonomers A and B, $\mathbf{k}_{\text{des}}$ can be expressed as

$$\mathbf{k}_{\text{des}} = k_{\text{des,A}}(n_A/\mathbf{n}) + k_{\text{des,B}}(n_B/\mathbf{n}) \tag{4.29}$$

where $k_{\text{des,A}}$ and $k_{\text{des,B}}$ are the free radical desorption rate coefficients for A-monomeric radical and B-monomeric radical, respectively, n_{A} and n_{B} are the average numbers of A radicals and B radicals per particle, respectively, and **n** is the total average number of free radicals per particle ($\mathbf{n} = n_{\text{A}} + n_{\text{B}}$). When all the desorbed monomeric radicals of A can be reabsorbed by the latex particles, $k_{\text{des,A}}$ can be calculated by the following equation:

$$k_{\text{des,A}} = k_{\text{des}}(\text{A})\{(k_{\text{tr,AA}} r_{\text{A}}[\text{M}_{\text{A}}]_p + k_{\text{tr,BA}}[\text{M}_{\text{B}}]_p)/[r_{\text{B}}[k_{\text{des}}(\text{A})\mathbf{n}/k_{p,\text{AA}}] + [\text{M}_{\text{A}}]_p]\} \tag{4.30}$$

where $k_{\text{des}}(\text{A})$ is the desorption rate constant for monomeric radicals of type A, as defined by Eq. (4.23). The parameters $k_{\text{tr,AA}}$ and $k_{\text{tr,BA}}$ are the rate coefficients for the chain transfer of free radicals terminated with a reactive unit of type A and free radicals terminated with a reactive unit of type B to monomer A, respectively, r_A and r_B are the reactivity ratios of monomers A and B, respectively, $[M_A]_p$ and $[M_B]_p$ are the concentrations of monomers A and B in the latex particles, respectively, and $k_{p,AA}$ is the rate constant for the propagation of free radicals terminated with a reactive unit of type A with monomer A.

Lopez et al. [63] investigated the seeded emulsion copolymerization of styrene and *n*-butyl acrylate. The reaction parameters chosen for this study include the size and concentration of seed latex particles and the concentration of initiator. A mechanistic model was used to simulate the emulsion copolymerization kinetics. The rate coefficient for the desorption of free radicals out of the latex poarticles developed by Forcada and Asua [64] was incorporated into the kinetic model. Barudio et al. [65] developed a mechanistic model for the emulsion copolymerization kinetics, in which the pseudo-homopolymerization approach [30, 52] was used to predict the free radical desorption process. Saldivar et al. [66] first reviewed the subject dealing with mathematical modeling of emulsion copolymerization systems and developed a comprehensive mechanistic model for predicting the related reaction kinetics. Extensive discussion on the average free radical desorption rate constant that is applicable to emulsion copolymerization systems can be found in this reference. Later, the validity of this kinetic model was justified by the experimental data [67, 68]. Barandiaran et al. [69] developed a method to estimate the free radical desorption rate constants for emulsion copolymerizations of binary monomers with the average number of free radicals per particle smaller than 0.5. The binary systems chosen for demonstration are the pairs of methyl acrylate–vinyl acetate ($\mathbf{n} < 0.5$) and methyl methacrylate–*n*-butyl acrylate ($\mathbf{n} > 0.5$).

4.3.3 Effect of Interfacial Properties on Desorption of Free Radicals

It is noteworthy that a basic assumption made in the derivation of the free radical desorption rate constant is that the adsorbed layer of surfactant or stabilizer surrounding the particle does not act as a barrier against the molecular diffusion of free radicals out of the particle. Nevertheless, a significant reduction (one order of magnitude) in the free radical desorption rate constant can happen in the emulsion polymerization of styrene stabilized by a polymeric surfactant [42]. This can be attributed to the steric barrier established by the adsorbed polymeric surfactant molecules on the particle surface, which retards the desorption of free radicals out of the particle. Coen et al. [70] studied the reaction kinetics of the seeded emulsion polymerization of styrene. The polystyrene seed latex particles were stabilized by the anionic random copolymer of styrene and acrylic acid. For reference, the polystyrene seed latex particles stabilized by a conventional anionic surfactant were also included in this study. The electrosteric effect of the latex particle surface layer containing the polyelectrolyte is the greatly reduced rate of desorption of free radicals out of the particle as compared to the counterpart associated with a simple anionic surfactant. A similar result was also reported for the seeded emulsion polymerization of styrene stabilized by polyacrylic acid [71].

4.4 GROWTH OF LATEX PARTICLES

After the completion of particle nucleation, the number of latex particles remains relatively constant with the progress of polymerization, provided that secondary nucleation of particle nuclei and flocculation of particles are absent from the reaction system (Smith–Ewart Interval II) [1, 72–76]. Immediately after the end of particle nucleation, monomer conversion is relatively low and a major proportion of monomer is present in the emulsified monomer droplets (>10° μm in diameter, 10^{12}–10^{14} dm^{-3} in number). These monomer droplets serve as a reservoir to supply the growing particle nuclei with monomer for free radical polymerization taking place therein.

In conventional emulsion polymerization, the rate of polymerization is dependent on the propagation rate constant, which is a function of the type of monomers and temperature, the concentration of monomer in the latex particles, the average number of free radicals per particle (Chapter 4, Sections 4.1–4.3), and the number of latex particles per unit volume of water (Chapter 3). These latex particles are the primary reaction loci, in which most monomer molecules are consumed via the propagation reaction with free radicals. The relatively hydrophobic monomer molecules residing in monomer droplets must diffuse through the continuous aqueous phase saturated with monomer and then penetrate into monomer-swollen polymer particles. The driving force for such mass transfer is the monomer concentration gradient established between the monomer droplet phase and the monomer-swollen polymer particle phase. Furthermore, the rate of polymerization increases linearly with increasing concentration of monomer in latex particles according to Eq. (4.1). The major factors that govern the concentration of monomer in the submicron reaction loci during polymerization are the foci of this section.

4.4.1 Thermodynamic Consideration

It is generally accepted that the number of latex particles per unit volume of water, the average number of free radicals per particle ($\mathbf{n}$ = 0.5), and the concentration of monomer in the particles are constant for emulsion polymerization systems that follow the ideal Smith–Ewart Case 2 kinetics. As a result, a constant reaction rate period can be observed during emulsion polymerization. Monomer molecules must be transferred from the gigantic monomer droplets to the growing submicron latex particles to supply the reaction. A dynamic balance between the rate of consumption of monomer in the latex particles and the rate of diffusion of monomer molecules from the monomer droplets to the particles may thus be established, and this results in a constant concentration of monomer in the major reaction loci during polymerization.

One potential method developed to predict the concentration of monomer in polymer particles in emulsion polymerization is the thermodynamic approach [77–80]. Morton et al. [77] proposed the following equation for calculating the equilibrium concentration of monomer in the polymer particles:

$$2\Phi_m\sigma/(RTr) = -[\ln(1-\Phi_p)+(1-1/X_n)\Phi_p+\chi\Phi_p^2] \qquad (4.31)$$

where Φ_m and Φ_p are the volume fractions of monomer and polymer, respectively, in the monomer-swollen polymer particles ($\Phi_m + \Phi_p = 1$), σ is the oil–water interfacial tension, R is the gas constant, T is the absolute temperature, r is the radius of latex particles, X_n is the number average degree of polymerization, and χ is the Flory–Huggins interaction parameter between monomer and polymer. It should be noted that the second term on the right-hand side of Eq. (4.31) can be neglected because the number average degree of polymerization is very large in typical emulsion polymerization systems. This equation was experimentally confirmed for the equilibrium swelling of various latex products. Decreasing the particle–water interfacial tension (σ), increasing the latex particle size (r), or increasing the temperature (T) result in an increase in the equilibrium monomer concentration in the particles (ϕ_m) according to Eq. (4.31). During Smith–Ewart Interval II, latex particles continue to grow in size at the expense of monomer droplets. The expanding surface area of these growing latex particles then results in a reduction in the particle surface coverage by surfactant and, consequently, causes the particle–water interfacial tension to increase during polymerization. The concentration of monomer in the latex particles may thus remain relatively constant if the effect of the increased particle size is counterbalanced coincidently by the effect of the increased particle–water interfacial tension.

Strictly speaking, any model based on the time-independent thermodynamics cannot be used to adequately predict the concentration of monomer in latex particles during Smith–Ewart Interval II. This is because the free radical polymerization of monomer in the discrete latex particles is governed by the simultaneous kinetic events such as the generation of free radicals in the continuous aqueous phase, the absorption of free radicals by the particles, the propagation of free radicals with monomer molecules in the particles, the bimolecular termination of free radicals in the particles, and the desorption of free radicals out of the particles. The equilibrium (or saturation) concentration of monomer in the growing latex particles may not be achieved if the rate of consumption of monomer in the major reaction loci is much faster than that of diffusion of monomer molecules from the monomer droplets to the reaction loci. Therefore, the equilibrium concentration of monomer in the latex particles represents an upper limit that is ultimately attainable in the course of polymerization. Nevertheless, the general validity of Smith–Ewart Case 2 kinetics has been confirmed experimentally in a variety of emulsion polymerization systems dealing with relatively hydrophobic monomers such as styrene and butadiene. This implies that diffusion of monomer from the monomer droplets to the growing latex particles during Smith–Ewart Interval II is not the rate-limiting step. Furthermore, the concentration of monomer in the latex particles is most likely equal to its saturation value unless a delicate balance between the rate of diffusion of monomer from the monomer droplets to the particles and that of propagation of free radicals with monomer in the reaction loci can be achieved in the emulsion polymerization system.

Equation (4.31) can be used to calculate Φ_m (or Φ_p), and the saturated concentration of monomer in latex particles ($[M]_p$) can be estimated by the following expression:

$$[M]_p = \Phi_m / V_m \tag{4.32}$$

where V_m is the partial molar volume of monomer in the particle phase.

For example, typical values of the saturated concentration of monomer in the latex particles for the emulsion polymerization of styrene and methyl methacrylate, respectively, are in the range 5–6 and 6–7 mol dm^{-3} [30, 81].

4.4.2 Concentrations of Comonomers in Emulsion Copolymerization Systems

For emulsion copolymerization, monomers may show different partitioning behavior among the emulsified monomer droplets, the latex particles, and the continuous aqueous phase. This can have a significant influence on the polymerization kinetics and the copolymer composition. The following empirical equation can be used to estimate the individual saturated concentration of comonomer i in latex particles ($[M_i]_p$) [30, 82–84]:

$$[M_i]_p = 1/(a_i + b_i / W_{i,d}) \tag{4.33}$$

where a_i and b_i are the adjustable constants for comonomer i and $W_{i,d}$ is the weight fraction of comonomer i in the monomer droplets. The validity of this empirical equation was demonstrated by the experimental data obtained from the emulsion copolymerization of styrene and methyl methacrylate, as shown in Figure 4.5 [30]. It was shown that the saturated concentrations of comonomers in latex particles are relatively insensitive to changes in the particle diameter (25–110 nm) and the copolymer composition. In addition, the weight fraction of comonomer i in latex particles was found to be approximately equal to that of comonomer i in monomer droplets.

Based on the extended equation developed by Ugelstad et al. [85], the thermodynamic approach was adopted by a number of papers to study the

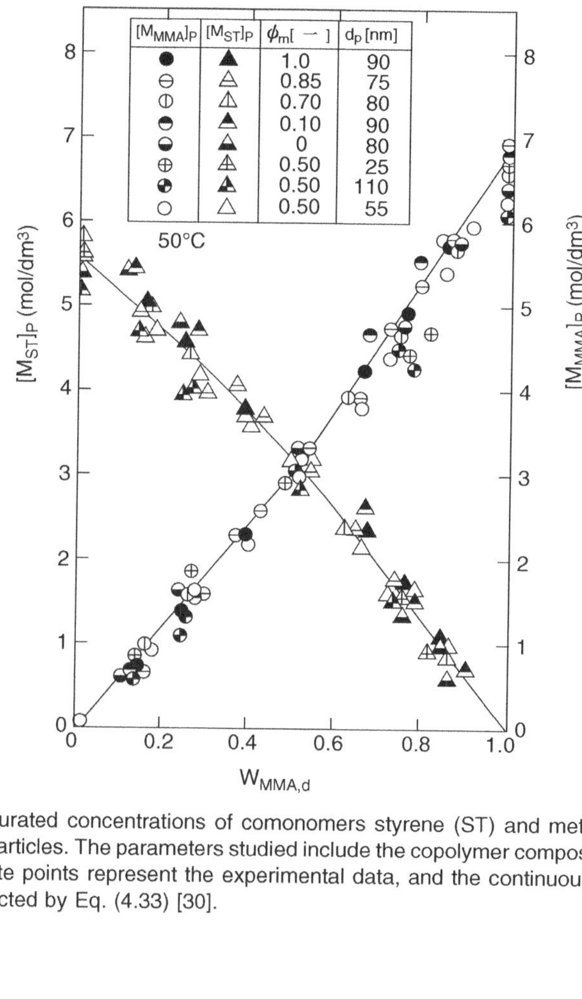

Figure 4.5. Saturated concentrations of comonomers styrene (ST) and methyl methacrylate (MMA) in latex particles. The parameters studied include the copolymer composition and particle size. The discrete points represent the experimental data, and the continuous lines represent the results predicted by Eq. (4.33) [30].

partitioning of comonomers among the emulsified monomer droplets, the latex particles and the continuous aqueous phase during emulsion copolymerization [86–97]. Considering the partial molar free energy of mixing of the individual comonomer with polymer in the latex particles, the contribution of the comonomer to the particle–water interfacial free energy, the partial molar free energy of the comonomer in monomer droplets, and the partial molar free energy of the comonomer in the continuous aqueous phase, the following equations were developed for comonomers i and j [87]:

$$\begin{aligned}
&\ln \Phi_{i,p} + (1 - m_{ij})\Phi_{j,p} + \Phi_p + \chi_{ij}\Phi_{j,p}^2 + \chi_{ip}\Phi_p^2 + \Phi_{j,p} + \Phi_p(\chi_{ij} + \chi_{ip} - \chi_{jp}m_{ij}) + 2V_{m,i}\gamma\Phi_p^{1/3}/(RTr) \\
&\quad = \ln \Phi_{i,d} + (1 - m_{ij})\Phi_{j,d} + \chi_{ij}\Phi_{j,d}^2 \\
&\quad = \ln([\mathrm{M}_i]_w/[\mathrm{M}_i]_{w,\mathrm{sat}}) \qquad (4.34)
\end{aligned}$$

$$\begin{aligned}
&\ln \Phi_{j,p} + (1 - m_{ji})\Phi_{i,p} + \Phi_p + \chi_{ij}\Phi_{i,p}^2 + \chi_{jp}\Phi_p^2 + \Phi_{i,p} \\
&\quad + \Phi_p(\chi_{ij} + \chi_{jp} - \chi_{ip}m_{ji}) + 2V_{m,j}\gamma\Phi_p^{1/3}/(RTr) \\
&\quad = \ln \Phi_{j,d} + (1 - m_{ji})\Phi_{i,d} + \chi_{ij}\Phi_{i,d}^2 \\
&\quad = \ln([\mathrm{M}_j]_w/[\mathrm{M}_j]_{w,\mathrm{sat}})
\end{aligned} \tag{4.35}$$

where Φ_p is the volume fraction of copolymer in latex particles, $\Phi_{i,p}$ and $\Phi_{j,p}$ are the volume fractions of comonomers i and j in polymer particles, respectively, and $\Phi_{i,d}$ and $\Phi_{j,d}$ are the volume fractions of comonomers i and j in monomer droplets, respectively. The parameters X_{ij}, X_{ip}, and X_{jp} are the Flory–Huggins interaction parameters between comonomers i and j, between comonomer i and polymer, and between comonomer j and polymer, respectively. The parameter m_{ij} (or m_{ji}) is the ratio of the molar volume of comonomer $i(V_{m,i})$ (or j $(V_{m,j})$) to that of comonomer j (or i). The symbols $[\mathrm{M}_i]_w$ and $[\mathrm{M}_j]_w$ are the concentrations of comonomers i and j in the continuous aqueous phase, respectively, whereas $[\mathrm{M}_i]_{w,\mathrm{sat}}$ and $[\mathrm{M}_j]_{w,\mathrm{sat}}$ represent the saturated concentrations of comonomers i and j in the continuous aqueous phase containing no other monomers, respectively.

It should be noted that derivation of Eq. (4.34) and (4.35) involves the reasonable assumption that the ratio of the molar volume of comonomer i (or j) to that of polymer is negligible compared to m_{ij} or m_{ji}. The following assumptions are used to further simplify the above two equations [87].

(a) The difference between the molar volumes of comonomers i and j is insignificant, and thus the ratio of the molar volume of comonomer i $(V_{m,i})$ (or j $(V_{m,j})$) to that of comonomer j (or i) is approximately equal to one.

(b) The contribution of the residual (e.g., enthalpic and nonconformational entropic) partial molar free energy of mixing of comonomers can be neglected in the monomer droplet phase.

(c) The Flory–Huggins interaction parameter χ_{ip} is equal to χ_{jp}.

The resultant expressions are as follows:

$$\Phi_{i,p}/\Phi_{j,p} = \Phi_{i,d}/\Phi_{j,d} \tag{4.36}$$

$$\Phi_{i,p} = \Phi_{i,d} \tag{4.37}$$

$$\Phi_{j,p} = \Phi_{j,d} \tag{4.38}$$

These equations are consistent with the work of Nomura and Fujita [82]. The validity of Eqs. (4.36)– (4.38) was confirmed experimentally for the emulsion copolymerizations of the styrene–methyl acrylate, styrene–n-butyl acrylate and methyl acrylate–n-butyl acrylate pairs. As a result of this study, the following empirical equations are recommended for the calculation of the concentrations of comonomers i ($[M_i]_p$) and j ($[M_j]_p$) in polymer particles:

$$[M_i]_p = \Phi_{i,d}\left[\left([M_i]_{p,\text{sat}} - [M_j]_{p,\text{sat}}\right)\Phi_{i,d} + [M_j]_{p,\text{sat}}\right] \tag{4.39}$$

$$[M_j]_p = \Phi_{j,d}\left[\left([M_j]_{p,\text{sat}} - [M_i]_{p,\text{sat}}\right)\Phi_{j,d} + [M_i]_{p,\text{sat}}\right] \tag{4.40}$$

where $[M_i]_{p,\text{sat}}$ and $[M_j]_{p,\text{sat}}$ are the saturated concentrations of comonomers i and j in polymer particles in the absence of other monomers, respectively.

The concentration of monomer in particle nuclei in the particle nucleation stage is generally assumed to be the saturated concentration involved in the Smith–Ewart Interval II. On the other hand, the concentration of monomer in latex particles in the absence of monomer droplets (Smith–Ewart Interval III) continues to decrease to the end of polymerization; the concentration of monomer in latex particles is linearly proportional to $(1 - X)$, where X is the fractional conversion of monomer. To minimize residual monomer in latex products is essential for the successful product development because of the potential hazard to end-users. An initiator pair of reducer and oxidant is usually post-added to the emulsion polymerization system to achieve this goal.

4.4.3 Competitive Growth of Latex Particles

Vanderhoff and Co-workers [98, 99] proposed the following empirical equation for the competitive growth of latex particles:

$$dv_p/dt = kd_p^c \tag{4.41}$$

where v_p is the volume of a single latex particle and k and c are the adjustable parameters. The values of c are equal to zero and 2.5, respectively, for the emulsion polymerizations with a water-soluble initiator (potassium persulfate) when the latex particle diameters are smaller than 150 nm and much larger than 150 nm, respectively. On the other hand, the value of c is equal to three when an oil-soluble initiator (benzoyl peroxide) is used to initiate the free radical polymerization. With the relationship $v_p = \pi d_p^3/6$, Eq. (4.41) can be transformed into the following equations:

$$dd_p/dt = (2k/\pi)d_p^{-2}, \qquad d_p < 150\,\text{nm} \tag{4.42}$$

$$dd_p/dt = (2k/\pi)d_p^{0.5}, \qquad d_p >> 150\,\text{nm} \tag{4.43}$$

The above equations indicate that small latex particles grow faster in size than larger ones, and therefore the resultant particle size distribution becomes narrower when d_p is smaller than 150 nm. By contrast, larger latex particles grow faster in size in comparison with small ones, provided that d_p is much larger than 150 nm.

Another popular equation available for predicting the volumetric growth rate of latex particles is shown below [100]:

$$dv_p/dt = k'[\mathrm{M}]_p\mathbf{n} \quad (4.44)$$

where k' is a constant closely related to the propagation reaction rate and the degree of swelling of polymer particles by monomer. Gardon [13] proposed that the rate of absorption of free radicals by latex particles is directly proportional to the total particle surface area. Van den Hul and Vanderhoff [101] showed that a significant fraction of polymer chain end-groups ($-SO_4^-$) are located at the latex particle surface. Sheinker and Medvedev [102] also suggested that polymeric radicals cannot penetrate the interior of latex particles due to the quite high viscosity. Consequently, the free radical polymerization takes place primarily near the latex particle surface layer. Chern and Poehlein [103] used a Monte Carlo simulation to illustrate that free radicals are not distributed uniformly within the latex particles for the emulsion polymerization initiated by a persulfate initiator. The volume concentration of free radical inside a latex particle is a decreasing exponential function from the surface toward the center of the particle. Brodnyan et al. [104] showed that the rate of polymerization is linearly proportional to the total latex particle surface area. Thus, the volumetric growth rate of latex particles should be proportional to the square of the particle diameter ($dv_p/dt \sim d_p^2$).

4.5 POLYMER MOLECULAR WEIGHT

Growth of polymeric radicals in latex particles is dependent on the propagation rate constant, the concentration of monomer in the particles and the average number of free radicals per particle. The number-average degree of polymerization (X_n) achieved in emulsion polymerization can reach 10^4. This is due to the effect of segregation of free radicals among a large population of latex particles. Formation of oligomeric radicals in the continuous aqueous phase does not contribute to the resultant polymer chain length significantly because the primary reaction loci are the discrete monomer-swollen polymer particles. The molecular weight and molecular weight distribution of emulsion polymer exhibit a significant influence on the mechanical and application properties (e.g., modulus, scrub resistance, solvent resistance, corrosion resistance, heat resistance, adhesion, and film formation).

In emulsion polymerization, X_n, in the absence of chain transfer reactions, is equal to the rate of growth of a polymeric radical divided by the rate of absorption of oligomeric radicals by the latex particle [105]:

$$X_n = k_p[\mathrm{M}]_p N_p/\rho_i \quad (4.45)$$

It should be noted that X_n is exactly the same as the average kinetic chain length in emulsion polymerization. This is simply because the termination reaction taking place in the monomer-swollen polymer particles involves a polymeric radical and an incoming oligomeric radical.

The pioneering work dealing with the molecular weight and molecular weight distribution of polymer produced in emulsion polymerization can be found in the literature [106, 107]. For example, the following expressions can be used to calculate the molecular weight and molecular weight distribution of polymer obtained from the Smith–Ewart Interval II in emulsion polymerization [106].

$$X_n = \beta(2\tau + 3 - e^{-2\tau})/(2\tau + 3 + e^{-2\tau}) \tag{4.46}$$

$$X_w/X_n = 2(2\tau + 3 + e^{-2\tau})(2\tau + 5/3 - 2e^{-2\tau} + e^{-2\tau})/(2\tau + 3 - e^{-2\tau})^2 \tag{4.47}$$

$$\tau = \rho_i t/n_p \tag{4.48}$$

$$\beta = k_p[\mathrm{M}]_p N_p/\rho_i \tag{4.49}$$

where X_w is the weight-average degree of polymerization. According to Eq. (4.46), the number-average degree of polymerization increases from $\beta/2$ to β during the Smith–Ewart Interval II. Equation (4.47) predicts that the polydispersity index (X_w/X_n) of the emulsion polymer molecular weight distribution increases from 4/3 to 2 (i.e., the molecular weight distribution becomes broader) with the progress of polymerization. Furthermore, Eq. (4.46) is simply reduced to Eq. (4.45) when τ is much larger than one (i.e., each latex particle contains a number of polymer chains). In general, increasing the number of latex particles per unit volume of water (i.e., increasing the degree of segregation of free radicals among the discrete reaction loci) and decreasing the concentration of initiator result in an increase in the molecular weight of emulsion polymer.

The Sydney group [108, 109] developed a theory for predicting the molecular weight distribution of linear polymer chains generated in emulsion polymerization. This mathematical model takes into account (a) the absorption of free radicals by latex particles, (b) the desorption of radicals out of the particles, (c) the bimolecular termination reaction, and (d) the chain transfer reaction. It was concluded that the compartmentalization of free radicals among the growing latex particles broadens the polymer molecular weight distribution and may even change the shape of the molecular weight distribution curve. Storti et al. [110, 111] adopted the approach of Markov processes and developed a kinetic model for predicting the instantaneous molecular weight distribution of polymer produced during emulsion polymerization. Another technique that is very useful in predicting the molecular weight and molecular weight distribution of polymer produced in emulsion polymerization is the Monte Carlo simulation method [112–122]. This unique technique can take into account any kinetic event (e.g., desorption of free radicals out of the latex particles) occurring in emulsion polymerization if its probability can be represented explicitly. In conventional Monte Carlo simulations of molecular build-up processes, monomeric units are added to each growing polymeric radical one-by-one. Thus, a multitude of random numbers and calculations are required to simulate the formation of each polymer chain. To get around this tedious problem, the competition technique was proposed to significantly reduce the computation time required for the Monte Carlo simulation [113, 116].

The application of these comprehensive models to the prediction of the emulsion polymer molecular weight distribution requires a fundamental understanding of the very complex reaction mechanisms and knowledge of various kinetic parameters (e.g., the rate coefficients for the absorption of free radicals by the latex particles, the desorption of radicals out of the particles, and the bimolecular termination reaction). However, these mathematical models in combination with the polymer molecular weight distribution data may serve as a useful tool for estimating the values of the kinetic parameters involved in emulsion polymerization.

A small amount of chain transfer agents such as very effective *n*-butyl mercaptan and *n*-dodecyl mercaptan are commonly used to reduce the molecular weights of emulsion polymers formed during the reaction. As aforementioned, incorporation of chain transfer agent into the emulsion polymerization system, which is heterogeneous in nature, promotes the formation of chain transfer agent radicals inside the latex particles. These mobile chain transfer agent radicals tend to desorb out of the particles, provided that they have some solubility in the continuous aqueous phase. Hence, the average number of free radicals per particle (or the rate of polymerization) is greatly reduced. For example, the water solubility of *n*-butyl mercaptan is higher than that of *n*-dodecyl mercaptan. As would be expected, *n*-butyl mercaptan has a stronger influence on the emulsion polymerization kinetics in comparison with the *n*-dodecyl mercaptan counterpart. It is also noteworthy that transport of the relatively hydrophobic chain transfer agent, *n*-dodecyl mercaptan, from the emulsified monomer droplets to the latex particles (reaction loci) is severely retarded. This transport phenomenon then limits the use of *n*-dodecyl mercaptan as an effective chain transfer agent in emulsion polymerization. On the other hand, crosslinking agents such as divinyl benzene and 1,6-hexanediol diacrylate are useful in significantly increasing the polymer molecular weight in emulsion polymerization. Alkali-soluble or -swellable resins obtained from emulsion copolymerizations of water-soluble monomers (e.g., acrylic acid and methacrylic acid) with relatively hydrophobic monomers (e.g., ethyl acrylate and styrene) are good examples of using chain transfer agents or crosslinking agents to control the molecular weight and molecular weight distribution and performance properties of emulsion polymers.

For those who are interested in the development of molecular weight and molecular weight distribution of polymer in emulsion polymerization in more detail, refer to reference 123.

REFERENCES

1. W. V. Smith and R. H. Ewart, *J. Chem. Phys.* **16**, 592 (1948).
2. R. N. Haward, *J. Polym. Sci.* **4**, 273 (1949).
3. W. H. Stockmayer, *J. Polym. Sci.* **24**, 313 (1957).
4. J. T. O'Toole, *J. Appl. Polym. Sci.* **9**, 1291 (1965).
5. N. Friis and A. E. Hamielec, *J. Polym. Sci., Polym. Chem. Ed.* **11**, 3321 (1973).
6. N. Friis and A. E. Hamielec, in *Emulsion Polymerization*, I. Piirma and J. L. Gardon (Eds.), ACS Symposium Series 24, American Chemical Society, Washington, D.C., 1976, p. 82.
7. D. C. Sundberg, J. Y. Hsieh, S. K. Soh, and R. F. Baldus, in *Emulsion Polymerization*, D. R. Bassett and A. E. Hamielec (Eds.), ACS Symposium Series 165, American Chemical Society, Washington, D.C., 1981, p. 327.
8. J. Ugelstad, P. C. Mork, and J. O. Aasen, *J. Polym. Sci., A-1* **5**, 2281 (1967).
9. J. Ugelstad and P. C. Mork, *Br. Polym. J.* **2**, 31 (1970).
10. J. Ugelstad, P. C. Mork, P. Dahl, and P. Rangnes, *J. Polym. Sci., Part C* **27**, 49 (1969).
11. M. Harada, M. Nomura, W. Eguchi, and S. Nagata, *J. Chem. Eng. Jpn.* **4**, 54 (1971).
12. N. Friis and L. Nyhagen, *J. Appl. Polym. Sci.* **17**, 2311 (1973).
13. J. L. Gardon, *J. Polym. Sci., Part A-1* **6**, 687 (1968).
14. B. S. Hawkett, D. H. Napper, and R. G. Gilbert, *J. Chem. Soc., Faraday Trans. 1* **76**, 1323 (1980).
15. S. W. Lansdowne, R. G. Gilbert, D. H. Napper, and D. F. Sangster, *J. Chem. Soc., Faraday Trans. 1* **76**, 1344 (1980).
16. D. C. Blackley, in *Emulsion Polymerization*, D. R. Bassett and A. E. Hamielec (Eds.), ACS Symposium Series 165, American Chemical Society, Washington, D.C., 1981, p. 437.
17. J. L. Gardon, *J. Polym. Sci., A-1* **6**, 623 (1968).
18. J. L. Gardon, *J. Polym. Sci., A-1* **6**, 643 (1968).
19. J. Ugelstad and F. K. Hansen, *Rubber Chem. Technol.* **49**, 536 (1976).
20. V. I. Yeliseeva, in *Emulsion Polymerization*, I. Piirma (Ed.), Academic Press, New York, 1982, p. 247.
21. I. A. Penboss, D. H. Napper, and R. G. Gilbert, *J. Chem. Soc., Faraday Trans. 1* **79**, 1257 (1983).
22. I. A. Maxwell, B. R. Morrison, D. H. Napper, and R. G. Gilbert, *Macromolecules* **24**, 1629 (1991).
23. G. L. Leslie, D. H. Napper, and R. G. Gilbert, *Aust. J. Chem.* **45**, 2057 (1992).
24. L. Lopez de Arbina, M. J. Barandiaran, L. M. Gugliotta, and J. M. Asua, *Polymer* **37**, 5907 (1996).
25. M. Harada, M. Nomura, H. Kojima, W. Eguchi, and S. Nagata, *J. Appl. Polym. Sci.* **16**, 811 (1972).
26. F. K. Hansen and J. Ugelstad, *J. Polym. Sci., Polym. Chem. Ed.* **16**, 1953 (1978).

27. F. K. Hansen, *Chem. Eng. Sci.* **48**, 437 (1993).
28. M. Nomura and M. Harada, *J. Appl. Polym. Sci.* **26**, 17 (1981).
29. M. Nomura, in *Emulsion Polymerization*, I. Piirma (Ed.), Academic Press, New York, 1982, p. 191.
30. M. Nomura, K. Yamamoto, I. Horie, K. Fujita, and M. Harada, *J. Appl. Polym. Sci.* **27**, 2483 (1982).
31. E. Unzueta and J. Forcada, *J. Appl. Polym. Sci.* **66**, 445 (1997).
32. C. Sayer, M. Palma, and R. Giudici, *Ind. Eng. Chem. Res.* **41**, 1733 (2002).
33. M. Nomura, M. Harada, W. Eguchi, and S. Nagata, in *Emulsion Polymerization*, I. Piirma and J. L. Gardon (Eds.), *ACS Symposium Series*, No. 24, Washington, D.C., 1976, p. 102.
34. P. H. H. Araujo, J. C. de la Cal, J. M. Asua, and J. C. Pinto, *Macromol. Theor. Simul.* **10**, 769 (2001).
35. Y. Dong and D. C. Sundberg, *Macromolecules* **35**, 8185 (2002).
36. H. A. S. Schoonbrood, A. L. German, and R. G. Gilbert, *Macromolecules* **28**, 34 (1995).
37. C. Marestin, A. Guyot, and J. Claverie, *Macromolecules* **31**, 1686 (1998).
38. D. Colombie, E. D. Sudol, and M. S. El-Aasser, *Macromolecules* **33**, 4347 (2000).
39. V. Liotta, C. Georgakis, E. D. Sudol, and M. S. El-Aasser, *Ind. Eng. Chem. Res.* **36**, 3252 (1997).
40. K. Tauer and R. Deckwer, *Acta Polym.* **49**, 411 (1998).
41. M. E. Adams, M. Trau, R. G. Gilbert, D. H. Napper, and D. F. Sangster, *Aust. J. Chem.* **41**, 1799 (1988).
42. J. M. H. Kusters, D. H. Napper, and R. G. Gilbert, *Macromolecules* **25**, 7043 (1992).
43. I. A. Penboss, R. G. Gilbert, and D. H. Napper, *J. Chem. Soc., Faraday Trans. 1* **82**, 2247 (1986).
44. R. G. Gilbert and D. H. Napper, *J. Macromol. Sci.-Rev. Macromol. Chem. Phys.* **23**, 127 (1983).
45. G. Lichti, D. F. Sangster, B. C. Whang, D. H. Napper, and R. G. Gilbert, *J. Chem. Soc., Faraday Trans. 1* **80**, 2911 (1984).
46. H. C. Lee and G. W. Poehlein, *Polym. Process Eng.* **5**, 37 (1987).
47. M. Nomura, M. Harada, K. Nakagawara, W. Eguchi, and S. Nagata, *J. Chem. Eng. Jpn.* **4**, 160 (1971).
48. H. C. Lee and G. W. Poehlein, *J. Disper. Sci. Technol.* **5**, 247 (1984).
49. H. C. Lee and G. W. Poehlein, *Chem. Eng. Sci.* **41**, 1023 (1986).
50. F. K. Hansen and J. Ugelstad, *J. Polym. Sci., Polym. Chem. Ed.* **17**, 3047 (1979).
51. F. K. Hansen and J. Ugelstad, *Makromol. Chem.* **180**, 2423 (1979).
52. M. Nomura, M. Kubo, and K. Fujita, *J. Appl. Polym. Sci.* **28**, 2767 (1983).
53. C. S. Chern, *J. Appl. Polym. Sci.* **56**, 231 (1995).
54. C. S. Chern and C. Lee, *J. Polym. Sci., Part A: Polym. Chem.* **40**, 1608 (2002).
55. R. G. Gilbert, *Emulsion Polymerization, A Mechanistic Approach*, Academic Press, London, 1995.

56. J. M. Asua, E. D. Sudol, and M. S. El-Aasser, *J. Polym. Sci., Polym. Chem. Ed.* **27**, 3903 (1989).
57. M. J. Barandiaran and J. M. Asua, *J. Polym. Sci., Polym. Chem. Ed.* **34**, 309 (1996).
58. J. M. Asua, *Polymer* **39**, 2061 (1998).
59. B. R. Morrison, B. S. Casey, I. Lacik, G. L. Leslie, D. F. Sangster, R. G. Gilbert, and D. H. Napper, *J. Polym. Sci., Polym. Chem. Ed.* **32**, 631 (1994).
60. J. U. Kim and H. H. Lee, *Polymer* **37**, 1941 (1996).
61. S. J. Fang, K. Wang, and Z. R. Pan, *Polymer* **44**, 1385 (2003).
62. W. V. Smith, *J. Am. Chem. Soc.* **68**, 2059 (1946).
63. L. Lopez de Arbina, M. J. Barandiaran, L. M. Gugliotta, and J. M. Asua, *Polymer* **38**, 143 (1997).
64. J. Forcada and J. M. Asua, *J. Polym. Sci., Polym. Chem. Ed.* **28**, 987 (1990).
65. I. Barudio, J. Guillot, and G. Fevotte, *J. Polym. Sci., Polym. Chem. Ed.* **36**, 157 (1998).
66. E. Saldivar, P. Dafniotis, and W. H. Ray, *J. Macromol. Sci., Rev. Macromol. Chem. Phys.* **38**, 207 (1998).
67. O. Araujo, R. Giudici, E. Saldivar, and W. H. Ray, *J. Appl. Polym. Sci.* **79**, 2360 (2001).
68. E. Saldivar, O. Araujo, R. Giudici, and C. Lopez-Barron, *J. Appl. Polym. Sci.* **79**, 2380 (2001).
69. M. J. Barandiaran, L. Lopez de Arbina, J. C. de la Cal, L. M. Gugliotta, and J. M. Asua, *J. Appl. Polym. Sci.* **55**, 231 (1995).
70. E. M. Coen, R. A. Lyons, and R. G. Gilbert, *Macromolecules* **29**, 5128 (1996).
71. L. Vorwerg and R. G. Gilbert, *Macromolecules* **33**, 6693 (2000).
72. W. D. Harkins, *J. Chem. Phys.* **13**, 381 (1945).
73. W. D. Harkins, *J. Chem. Phys.* **14**, 47 (1946).
74. W. D. Harkins, *J. Am. Chem. Soc.* **69**, 1428 (1947).
75. W. V. Smith, *J. Am. Chem. Soc.* **70**, 3695 (1948).
76. W. V. Smith, *J. Am. Chem. Soc.* **71**, 4077 (1949).
77. M. Morton, S. Kaizerman, and M. W. Altier, *J. Colloid Sci.* **9**, 300 (1954).
78. J. L. Gardon, *J. Polym. Sci., A-1* **6**, 2859 (1968).
79. I. A. Maxwell, J. Kurja, G. H. J. van Doremaele, A. L. German, and B. R. Morrison, *Makromol. Chem.* **193**, 2049 (1992).
80. M. Antonietti, H. Kasper, and K. Tauer, *Langmuir* **12**, 6211 (1996).
81. E. Bartholome, H. Gerrens, R. Herbeck, and H. M. Weitz, *Z. Elektrochem.* **60**, 334 (1956).
82. M. Nomura and K. Fujita, *Makromol. Chem. Suppl.* **10/11**, 25 (1985).
83. M. Nomura, *Kobunshi Kagaku* **36**, 680 (1987).
84. M. Nomura, I. Horie, M. Kubo, and K. Fujita, *J. Appl. Polym. Sci.* **37**, 1029 (1989).
85. J. Ugelstad, P. C. Mork, H. R. Mfutakamba, E. Soleimany, I. Nordhuus, R. Schmid, A. Berge, T. Ellingsen, O. Aune, and K. Nustad, In *Science and Technology of*

Polymer Colloids, Vol. 1 (NATO ASI Ser. E:67), Martinus Nijhoff, Boston, MA, 1983, p. 51.

86. P. Mathey and J. Guillot, *Polymer* **32**, 934 (1991).
87. I. A. Maxwell, J. Kurja, G. H. J. van Doremaele, and A. L. German, *Makromol. Chem.* **193**, 2065 (1992).
88. A. M. Aerdts, M. W. A. Boei, and A. L. German, *Polymer* **34**, 574 (1993).
89. L. F. J. Noel, I. A. Maxwell, and A. L. German, *Macromolecules* **26**. 2911 (1993).
90. H. A. S. Schoonbrood and A. L. German, *Macromol. Rapid Commun.* **15**, 259 (1994).
91. L. F. J. Noel, J. M. A. M. van Zon, I. A. Maxwell, and A. L. German, *J. Polym. Sci., Polym. Chem. Ed.* **32**, 1009 (1994).
92. H. A. S. Schoonbrood, M. A. T. V. D. Boom, A. L. German, and J. Hutovic, *J. Polym. Sci., Polym. Chem. Ed.* **32**, 2311 (1994).
93. L. M. Gugliotta, G. Arzamendi, and J. M. Asua, *J. Appl. Polym. Sci.* **55**, 1017 (1995).
94. M. Nomura, X. Liu, K. Ishitani, and K. Fujita, *J. Polym. Sci., Polym. Phys. Ed.* **32**, 2491 (1994).
95. X. Liu, M. Nomura, and K. Fujita, *J. Appl. Polym. Sci.* **64**, 931 (1997).
96. X. Liu, M. Nomura, Y. H. Liu, K. Ishitani, and K. Fujita, *Ind. Eng. Chem. Res.* **36**, 1218 (1997).
97. O. Karlsson and B. Wesslen, *J. Appl. Polym. Sci.* **70**, 2041 (1998).
98. E. B. Bradford, J. W. Vanderhoff, and T. Alfrey Jr., *J. Colloid Sci.* **11**, 135 (1956).
99. J. W. Vanderhoff, J. F. Vitkuske, E. B. Bradford, and T. AlfreyJr., *J. Polym. Sci.* **20**, 225 (1956).
100. G. W. Poehlein and J. W. Vanderhoff, *J. Polym. Sci.* **11**, 447 (1973).
101. H. J. van den Hul and J. W. Vanderhoff, *Br. Polym. J.* **2**, 121 (1970).
102. A. P. Shcinker and S. S. Medvedev, *Rubber Chem. Technol.* **29**, 121 (1956).
103. C. S. Chern and G. W. Poehlein, *J. Polym. Sci., Polym. Chem. Ed.* **25**, 617 (1987).
104. J. G. Brodnyan, J. A. Cala, T. Konen, and E. L. Kelley, *J. Colloid Sci.* **18**, 73 (1963).
105. V. I. Eliseeva, S. S. Ivanchev, S. I. Kuchanov, and A. V. Lebedev, *Emulsion Polymerization and Its Application in Industry*, Consultants Bureau, New York, 1981.
106. G. M. Saidel and S. Katz, *J. Polym. Sci., Part C* **27**, 149 (1969).
107. D. C. Sundberg and J. D. Eliassen, in *Polymer Colloids*, R. M. Fitch (Ed.), Plenum Press, New York, 1971, p. 153.
108. G. Lichti, R. G. Gilbert, and D. H. Napper, in *Polymer Colloids II*, R. M. Fitch (Ed.), Plenum Press, New York, 1980, p. 651.
109. D. H. Napper, G. Lichti, and R. G. Gilbert, in *Emulsion Polymers and Emulsion Polymerization*, D. R. Bassett and A. E. Hamielec (Eds.), ACS Symposium Series, No. 165, Washington, D.C., 1981, p. 105.
110. G. Storti, G. Polotti, M. Cociani, and M. Morbidelli, *J. Polym. Sci., Polym. Chem. Ed.* **30**, 731 (1992).

111. G. Storti, G. Polotti, P. Canu, and M. Morbidelli, *J. Polym. Sci., Polym. Chem. Ed.* **30**, 751 (1992).
112. H. Tobita and K. Yamamoto, *Macromolecules* **27**, 3389 (1994).
113. H. Tobita, Y. Takada, and M. Nomura, *Macromolecules* **27**, 3804 (1994).
114. H. Tobita, *Polymer* **35**, 3023 (1994).
115. H. Tobita, *Polymer* **35**, 3032 (1994).
116. H. Tobita, Y. Takada, and M. Nomura, *J. Polym. Sci., Polym. Phys. Ed.* **33**, 441 (1995).
117. H. Tobita, *Macromolecules* **28**, 5128 (1995).
118. H. Tobita, *Acta Polym.* **46**, 185 (1995).
119. H. Tobita and Y. Uemura, *J. Polym. Sci., Polym. Phys. Ed.* **34**, 1403 (1996).
120. H. Tobita and Y. Yoshihara, *J. Polym. Sci., Polym. Phys. Ed.* **34**, 1415 (1996).
121. H. Tobita, *J. Polym. Sci., Polym. Phys. Ed.* **35**, 1515 (1997).
122. H. Tobita and M. Nomura, *Colloid Surface A* **153**, 119 (1999).
123. M. Nomura, H. Tobita, and K. Suzuki, *Adv. Polym. Sci.* **175**, 1 (2005).

5

MINIEMULSION POLYMERIZATION

Generally, emulsified monomer droplets (~1–10 μm in diameter, 10^{12}–10^{14} dm^{-3} in number density) are not considered to significantly contribute to the particle nucleation process to any appreciable extent in conventional emulsion polymerization. This is simply because the total monomer droplet surface area is so small that monomer droplets are ineffective in competing with monomer-swollen micelles (~0.05–1 μm in diameter, 10^{19}–10^{21} dm^{-3} in number density) for the incoming oligomeric radicals. However, homogenized submicron monomer droplets containing an extremely hydrophobic, low-molecular-weight compound such as hexadecane, cetyl alcohol, or a small amount of dissolved polymer may become the predominant particle nucleation loci (monomer droplet nucleation). This polymerization technique is termed *miniemulsion polymerization* [1–13].

One unique characteristic of miniemulsion polymerization is that hydrophobic compounds can be incorporated into the interior of polymer particles during the reaction, and this cannot be achieved by conventional emulsion polymerization because strongly hydrophobic compounds are incapable of diffusing from monomer droplets, across the continuous aqueous phase, and into latex particles (reaction loci). The extremely hydrophobic, low-molecular-weight compound is often termed the cosurfactant in the literature, even though it may not exhibit any surfactant property in the preparation of miniemulsion (e.g., hexadecane). Perhaps a more appropriate terminology for such compounds is *costabilizer*. Such compounds stabilize submicron monomer droplets by reducing the thermodynamic driving force for the transport of monomer. Such transport would result in an increase in

Principles and Applications of Emulsion Polymerization, by Chorng-Shyan Chern

the concentration of the hydrophobe in the monomer droplets, a free energy cost.

5.1 POLYMERIZATION IN MONOMER DROPLETS [14]

The formation of latex particles resulting from free radical polymerization within emulsified monomer droplets is often ignored for many emulsion polymerization systems. This position is usually justified because the amount of polymerization in the monomer droplets is often less than 1% of the total, and the ratio of the number of latex particles originating from monomer droplets to the total number of latex particles is extremely small. Nevertheless, there is no barrier that would prevent oligomeric radicals from entering the monomer droplets and then initiating polymerization therein. When this happens, the monomer droplet becomes a dilute polymer–monomer solution and the thermodynamic driving force still results in the transport of monomer molecules from the droplets to the latex particles that have a higher polymer concentration. This transport phenomenon continues until the relative concentrations in both types of monomer-swollen polymer particles are similar. Thus, for example, a monomer droplet with a diameter of 10 μm may become a latex particle with a diameter of 1 μm—a loss of 99+% of its original monomer to provide for growth of the other, more numerous particles that have been formed by other mechanisms such as micellar nucleation and homogeneous nucleation. Furthermore, the presence of hydrophobic polymer in monomer droplets significantly retards the degradation of droplets; therefore, these droplets may even survive well beyond Smith–Ewart Interval II. Nevertheless, the shrinking monomer droplets with polymer inside will gradually lose their characteristics with the progress of polymerization and eventually become indistinguishable, except perhaps for their possible larger size from those latex particles generated via other mechanisms.

While the extent of polymerization in monomer droplets is often very small, the following reaction conditions and/or recipes can promote polymer reactions in the droplets.

(a) Intense emulsification results in a larger population of smaller monomer droplets with a larger total droplet surface area. Thus, less surfactant is available to form monomer-swollen micelles and/or to stabilize nucleated latex particles. Miniemulsion polymerization represents the extreme limit of polymerization in monomer droplets. Special stabilizer systems and intense homogenization generate numerous very small monomer droplets and particle formation can be shifted almost completely to those droplets. Bimodal particle size distributions can be achieved if this shift is not complete.

(b) Relatively slow diffusion of monomer molecules from the monomer droplets also increases the magnitude of polymerization therein. Even in conventional recipes, the monomer droplets remain long after Smith–Ewart Interval I and are potential sites for capture of oligomeric radicals. This phenomenon is magnified with fast polymer reactions (i.e., less time for monomer diffusion) and if monomers with water solubility lower than that of styrene, for example, are used. In fact, highly hydrophobic monomers cannot be effectively polymerized in conventional emulsion polymerization processes. In this case, miniemulsion polymerization is the process of choice for the production of polymer colloids.

In summary, formation of particle nuclei from emulsified monomer droplets is almost certain to occur in any emulsion polymerization system in which these droplets are present. As mentioned earlier, however, monomer droplets containing polymer will primarily serve as reservoirs to provide monomer to the much more numerous and smaller latex particles formed by other particle nucleation mechanisms. Polymerization in monomer droplets can be eliminated or at least minimized by using seed polymer particles and slowly adding monomer (neat or as an emulsion) to supply the growing seed particles (i.e., seeded semibatch emulsion polymerization under the monomer-starved condition).

5.2 STABILITY OF MONOMER EMULSIONS

5.2.1 Ostwald Ripening Effect

Satisfactory monomer droplet stability during storage or particle nucleation is a basic requirement for successful miniemulsion polymerization. In addition to the often observed flocculation and coalescence of monomer droplets, which are beyond the scope of this chapter, another emulsion degradation mechanism is the Ostwald ripening phenomenon. The solubility of monomer in the continuous aqueous phase increases exponentially with decreasing monomer droplet diameter, according to the following equation originally proposed by Kelvin [15]:

$$C_m(d_d) = C_m(\infty)\exp[4\sigma V_m/(RTd_d)] \tag{5.1}$$

where $C_m(d_d)$ and $C_m(\infty)$ are the solubility of monomer in the aqueous phase for monomer droplets with a diameter of d_d and the solubility of the bulk monomer in water, respectively, σ is the droplet–water interfacial tension, V_m is the molar volume of monomer in the droplets, R is the gas constant, and T is the absolute temperature. Such a chemical potential effect arising from those monomer droplets with different radii of curvature will allow monomer molecules in the smaller droplets to dissolve in water, diffuse through the aqueous phase, and then enter the larger droplets. Thus, larger monomer droplets tend to grow in size at the expense of smaller droplets and, ultimately, such a diffusional degradation process will destabilize miniemulsion products (termed the Ostwald ripening effect). Equation (5.1) also predicts that an increase in the monomer droplet–water interfacial tension (σ) via the adsorption of less surfactant molecules on the droplet surfaces, a decrease in the thermal energy (RT), or an increase in the solubility of the bulk monomer in water results in the increased solubility of monomer in the aqueous phase. Under these circumstances, the stronger Ostwald ripening effect makes it more difficult for the emulsion system to achieve satisfactory colloidal stability.

According to the Lifshitz–Slezov–Wagner (LSW) theory [15–18], the rate of Ostwald ripening (R_O) can be expressed as

$$\begin{aligned} R_O &= d(d_d^3)/dt \\ &= 64\sigma D_m V_m C_m(\infty)/(9RT) \end{aligned} \tag{5.2}$$

where D_m is the molecular diffusivity of monomer in water. Equation (5.2) predicts that the rate of Ostwald ripening increases linearly with increasing the solubility of the bulk monomer in water. The validity of this equation was confirmed by the experimental results obtained from measurements of Ostwald ripening rate for a series of emulsions comprising n-alkane (C_nH_{2n+2}, n = 9–16) with different water solubility stabilized by sodium dodecyl sulfate, as shown in Figure 5.1 [19].

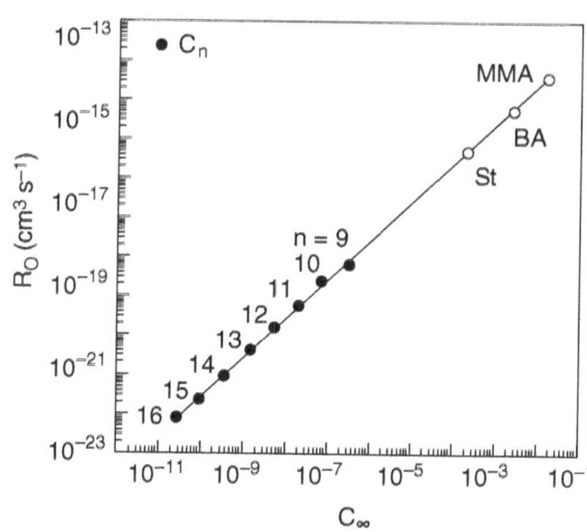

Figure 5.1. Rate of Ostwald ripening for emulsions as a function of the solubility of the constituent in water. The constituents of the oil phase include *n*-alkanes (*n* = 9–16) [19] and some common monomers. St, BA, and MMA represent styrene, *n*-butyl acrylate, and methyl methacrylate, respectively. The data of the solubility of monomers in water were used to estimate the Ostwald ripening rate of the homogenized monomer droplets via the extrapolation method.

5.2.2 Role of Costabilizer in Stabilizing Monomer Emulsions

Incorporation of 1–5 wt% costabilizer into the colloidal system can effectively retard the diffusion of monomer molecules from small monomer droplets to large ones due to the osmotic pressure effect [15, 20]. Diffusion of monomer species from a small monomer droplet to a large droplet results in a concentration gradient for costabilizer between these two droplets. However, unlike common monomers (e.g., styrene and methyl methacrylate), the extremely hydrophobic costabilizer molecules in the small monomer droplet are incapable of being dissolved in water, diffusing across the continuous aqueous phase, and then entering the large droplet. Thus, monomer molecules in the large monomer droplet are forced to migrate back to the small droplet in order to relax the concentration gradient for costabilizer established between these two droplets and a relatively stable miniemulsion product is obtained (Figure 5.2).

Based on the extended LSW theory, the rate of Ostwald ripening for the costabilizer containing miniemulsion can be calculated by the following equation [15]:

$$\begin{aligned} R_O &= d(d_d^3)/dt \\ &= 64\sigma D_c V_m C_c(\infty)/(9RT\phi_c) \end{aligned} \tag{5.3}$$

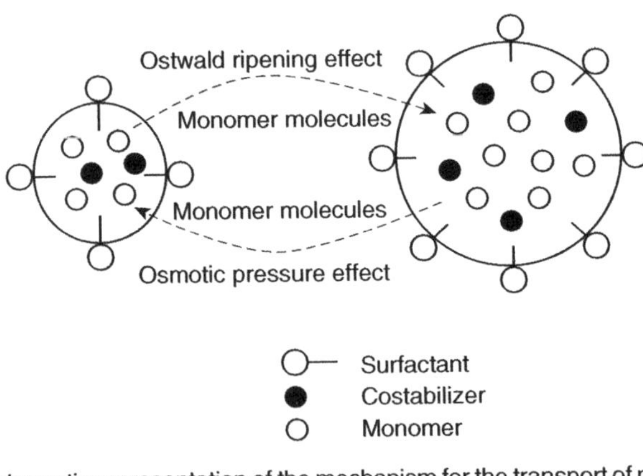

Figure 5.2. A schematic representation of the mechanism for the transport of monomer between a small monomer droplet and a large droplet. Monomer molecules tend to diffuse from the small monomer droplet to the large droplet due to the Ostwald ripening effect. This will cause a concentration gradient for costabilizer between these two monomer droplets. However, the very hydrophobic costabilizer in the small monomer droplet cannot be dissolved in water, diffuse across the continuous aqueous phase, and then enter the large droplet. Thus, monomer molecules in the large monomer droplet are forced to migrate back to the small droplet in order to relax the concentration gradient for costabilizer (termed the osmotic pressure effect), and a relatively stable miniemulsion product is obtained.

where D_c is the molecular diffusivity of costabilizer in water, $C_c(\infty)$ is the solubility of the bulk costabilizer in water, and ϕ_c is the volume fraction of costabilizer in the monomer droplets. Lowering the oil–water interfacial tension, decreasing the solubility of the bulk costabilizer in water, and increasing the level of costabilizer within the monomer droplets greatly enhance the stability of miniemulsion against the diffusional degradation. These effects form the theoretical basis of potential free radical polymerization of common monomers such as styrene and methyl methacrylate in the homogenized monomer droplets containing costabilizer (or hydrophobe).

Neglecting the effects of molar volume and interfacial tension, the following empirical equation can be used to reasonably predict the overall rate of Ostwald ripening for the monomer emulsion in the presence of costabilizer [18]:

$$R_O = [(\phi_m / R_{O,m}) + (\phi_c / R_{O,c})]^{-1} \tag{5.4}$$

where ϕ_m is the volume fraction of monomer in the monomer droplets and $R_{O,m}$ and $R_{O,c}$ represent the rates of Ostwald ripening for the single-component monomer emulsion and costabilizer emulsion, respectively. As a limiting case, the first term in the denominator can be neglected if $R_{O,c}$ is much smaller than $R_{O,m}$ (i.e., $\phi_m/R_{O,m} << \phi_c/R_{O,c}$). In this case, the overall rate of Ostwald ripening is primarily controlled by the hydrophobicity and amount of costabilizer present in the monomer droplets.

5.3 TYPE OF COSTABILIZERS IN MINIEMULSION POLYMERIZATION

In addition to the conventional long-chain alkanes (e.g., hexadecane) and alcohols (e.g., cetyl alcohol), the hydrophobic species that have been evaluated as costabilizer in the preparation of miniemulsion include polymers [21–27], oil-soluble initiators [28, 29], chain transfer agent [30, 31], dye [32], and reactive costabilizers [33–39]. In general, the less effective polymeric costabilizers do not generate stable monomer miniemulsion products, but they can retard the Ostwald ripening process to such an extent that nucleation in the monomer droplets can be achieved during polymerization. Miller et al. [22], for example, demonstrated the feasibility of preparing kinetically stable styrene miniemulsions using 1 wt% polystyrene as the costabilizer. It should be noted that this kind of miniemulsion in the absence of mechanical mixing creams rather rapidly upon aging.

Alduncin et al. [28] studied the effect of a series of initiators with different water solubility (lauroyl peroxide, benzoyl peroxide, and 2,2′-azobisisobutyronitrile) on styrene emulsion polymerization. It was concluded that, among these oil-soluble initiators, only lauroyl peroxide with the lowest water solubility (2×10^{-9} g per 100 g water) is hydrophobic enough to stabilize the homogenized monomer emulsion against the degradation of monomer droplets by the molecular diffusion process. The unique feature of this approach is that lauroyl peroxide not only acts as a costabilizer in stabilizing the monomer droplets, but also participates in the subsequent free radical polymerization. The initiator radical fragments generated by the thermal decomposition reaction eventually become part of the resultant polymer chains during the styrene miniemulsion polymerization.

The chain transfer agent, *n*-dodecyl mercaptan, was shown to be a quite effective costabilizer in preparing methyl methacrylate miniemulsions [30] or styrene miniemulsions [31]. For example, the shelf life of monomer miniemulsions comprising the relatively hydrophobic styrene ranges from 17 hours to 3 months. In miniemulsion polymerization, chain transfer agent species compete effectively with monomer molecules for free radicals to form chain transfer agent radicals inside the latex particles. Thus, similar to the idea of using oil-soluble initiators (e.g., lauroyl peroxide) as the costabilizer, the subsequent reinitiation reaction of chain transfer agent radicals with monomer molecules allows these radicals to be incorporated into the propagating polymer chains. The influence of the chain transfer agent costabilizer on the rate of polymerization is expected to be insignificant. This is simply because it is unlikely for the extremely hydrophobic chain transfer agent radicals to desorb from the latex particles and then reduce the average number of free radicals per particle. As expected, the molecular weight of polymer obtained from the miniemulsion polymerization with chain transfer agent as the costabilizer can be quite low.

Chern et al. [33–39] used stearyl methacrylate or lauryl methacrylate as the reactive costabilizer to stabilize styrene miniemulsion polymerizations. Just like conventional costabilizers (e.g., hexadecane), long-chain alkyl methacrylates act as costabilizers in stabilizing the homogenized submicron monomer droplets. Furthermore, the methacrylate group (—C=C(CH_3)COO—) of the polymerizable costabilizer can be chemically incorporated into latex particles in the subsequent free radical polymerization and thereby reduce the level of volatile organic compounds (VOC). As the polymerization proceeds, the reactive costabilizer concentration in the nucleated monomer droplets will decrease. The initial decrease of the costabilizer concentration should not cause any diffusional degradation because the hydrophobic polymer formed inside the nucleated monomer droplets can help stabilize the polymerizing miniemulsion.

Those who are interested in the recipes (e.g., monomers, surfactants, costabilizers, and initiators), preparation (e.g., homogenization equipments and processes), and characterization (e.g., monomer droplet size and droplet size distribution and colloidal stability) of monomer miniemulsions are referred to references 12 and 13.

5.4 MINIEMULSION POLYMERIZATION MECHANISMS AND KINETICS

5.4.1 Initial Conditions for Miniemulsion Polymerization Systems

The initial conditions for typical monomer miniemulsions are schematically shown in Figure 5.3. At a relatively low level of surfactant (e.g., lower than the critical micelle concentration), miniemulsions only consist of monomer droplets with a relatively broad droplet size distribution (Figure 5.3a). In this case, micellar nucleation can be ruled out. On the other hand, at a relatively high level of surfactant, in addition to monomer droplets (<10° μm in diameter), monomer-swollen micelles (~10° nm in diameter) may also exist in the polymerization system (Figure 5.3b). The ratio of the number of monomer droplets to that of monomer-swollen micelles is primarily controlled by the level of surfactant used to stabilize the colloidal system and the average droplet size and droplet size distribution. These factors determine the magnitude of the driving force (i.e., the amount of surfactant available) for the formation of monomer-swollen micelles. Under such circumstances, any particle nucleation mechanism (monomer droplet nucleation, micellar nucleation or homogeneous nucleation) can take place during polymerization.

The extremely hydrophobic costabilizer molecules are primarily present in the homogenized monomer droplets, and their concentration in the continuous aqueous phase is generally negligible. Initiators used in miniemulsion polymerization can be either water-soluble or oil-soluble. The loci in which

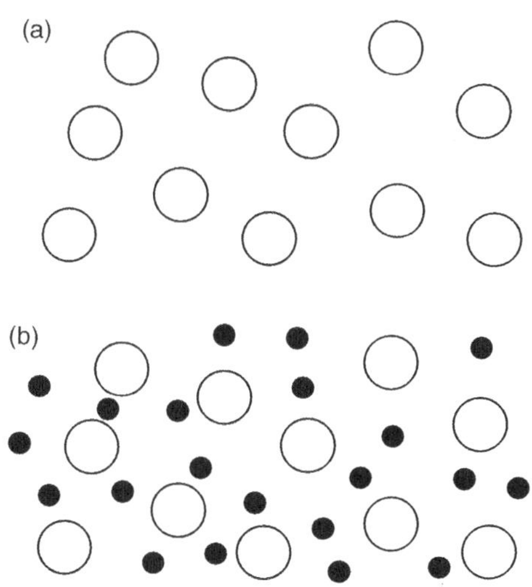

Figure 5.3. A schematic representation of the initial condition for the miniemulsion polymerization stabilized by surfactant below its critical micelle concentration **(a)** and above its critical micelle concentration **(b)**. The symbols ○ (<10^3 nm in diameter for miniemulsion polymerization) and ● (~10° nm in diameter) represent the homogenized monomer droplets and monomer-swollen micelles, respectively.

initiator molecules reside should have a significant influence on the particle nucleation process. Water-soluble initiator molecules present in the continuous aqueous phase are capable of not only entering the monomer droplets and monomer-swollen micelles (if present) to form particle nuclei but also promoting homogeneous nucleation, especially for those polymerization systems stabilized by very high levels of surfactant. On the other hand, oil-soluble initiator species experience difficulty in diffusing from the monomer droplets and monomer-swollen micelles (if present) into the aqueous phase; therefore, homogeneous nucleation is greatly retarded. The subject of competitive particle nucleation mechanisms in miniemulsion polymerization is one of the major foci of this section.

5.4.2 Particle Nucleation Mechanisms

The most important characteristic of miniemulsion polymerization is the transformation of the homogenized monomer droplets into latex particles via the capture of free radicals when a water-soluble initiator such as the persulfate initiator is used to initiate the free radical polymer reactions. However, this feature does not necessarily guarantee that the particle nucleation mechanisms other than monomer droplet nucleation can be ruled out. As will be shown later, previous studies dealing with nucleation of particle nuclei in miniemulsion polymerization often resulted in controversial conclusions. This subject is still open to discussion, and it represents a great challenge to polymer scientists.

As an extreme, in ideal miniemulsion polymerization, formation of latex particles directly follows the route of one-to-one copy of the homogenized monomer droplets, as evidenced by small-angle neutron scattering, conductivity, and surface tension measurements [40]. In other words, all the monomer droplets can be successfully converted into latex particles during polymerization, and hence the preservation of the original colloidal particle identity can be achieved. On the other hand, some studies suggest that only a fraction of monomer droplets are nucleated in the styrene or vinyl acetate-*n*-butyl acrylate miniemulsion polymerization [5, 6]. This confusing situation may be, to a large extent, due to the lack of reliable characterization methods for the homogenized monomer droplets and the inadequate techniques developed to compare the resultant latex particles with the droplets initially present in the polymerization system [12, 13]. This is because size and size distributions of the original monomer droplets very similar to those of the resultant latex particles have been often taken as the supporting evidence for the preservation of the original colloidal particle identity in miniemulsion polymerization. The polymer particle size and particle size distribution of latex products can be characterized with well-established techniques such as electron microscopy and dynamic light scattering. However, determination of the initial monomer droplet size and droplet size distribution represents a very challenging task to colloidal scientists. Some potential techniques developed for this purpose include the small-angle neutron scattering [40] and the indirect method that involves determination of the critical micelle concentration of a particular miniemulsion to calculate the oil–water interfacial area and, therefore, the droplet size [41]. In the author's opinion, the colloidal particle size and particle size distribution data obtained from a particular monomer miniemulsion and its corresponding latex product alone are insufficient to confirm or deny the predominant nucleation in the monomer droplets.

The following summarizes a general reaction scheme for the formation of particle nuclei in miniemulsion polymerization [13]:

(a) Formation of initiator radicals in the continuous aqueous phase via the thermal decomposition of a water-soluble initiator such as the persulfate initiator.
(b) Propagation of initiator radicals with monomer molecules in the aqueous phase. The hydrophobicity of oligomeric radicals (i.e., the tendency for these free radicals to diffuse toward a hydrophobic environment) increases with increasing the free radical chain length.
(c) After a critical chain length is reached, oligomeric radicals start to enter into the monomer droplets (monomer droplet nucleation) or monomer-swollen micelles (micellar nucleation).

Ugelstad et al. [1] first reported nucleation and polymerization in the homogenized styrene droplets with a diameter of smaller than 0.7 μm achieved by mechanical agitation. This unique polymerization technique resulted in stable latex products with a broad particle size distribution. This can be attributed to the ineffective homogenization method used in preparing styrene miniemulsions, according to the comments of Antonietti and Landfester [12]. Sodium dodecyl sulfate and cetyl alcohol were used as the surfactant and costabilizer, respectively. Formation of particle nuclei in an ideal miniemulsion polymerization system is governed by the monomer droplet nucleation mechanism. Every single monomer droplet should be nucleated due to the statistics of the absorption of free radicals by the droplets and the overall droplet size [12]. As a result, the number of latex particles remains relatively constant throughout the polymerization. Predominant nucleation in the monomer droplets was supported by the work of Reimers and Schork [26, 27, 30] and Landfester et al. [40, 42]. On the other hand, Choi et al. [5] studied the styrene miniemulsion polymerization using cetyl alcohol as the costabilizer and observed that only 20% of the initial monomer droplets were successfully transformed into latex particles.

Chern et al. [32, 37–39] used a water-insoluble dye as the molecular probe for the particle nucleation loci in styrene miniemulsion polymerizations stabilized by a surfactant concentration lower than its critical micelle concentration.

Lauryl methacrylate or stearyl methacrylate was used as the costabilizer to retard the Ostwald ripening effect. First, a conventional emulsion polymerization was carried out to investigate the mass transfer of dye molecules, which were added to the reaction mixture at a monomer conversion of 32.4%, from the dye bulk phase to the growing latex particles. The experimental data show that transport of dye species from the bulk phase, across the continuous aqueous phase, and then into the reaction loci (i.e., latex particles) is insignificant due to the very low solubility of the dye in water. This result also eliminates the possibility of forming another population of monomer droplets incorporated with dye via the diffusion of dye molecules from latex particles into the continuous aqueous phase. This is because the extremely hydrophobic dye molecules are content with being buried inside the latex particles and, perhaps, the diffusion coefficient of dye with a molecular weight of 10^3 in a highly viscous polymeric matrix is not large enough to allow the desorption process to occur. A mass balance was then established to determine the number of latex particles per unit volume of water originating from monomer droplet nucleation and the number of primary particles per unit volume of water generated in the continuous aqueous phase. The accuracy of this method relies on producing a stable monomer miniemulsion during polymerization. The experimental data thus obtained from this series of studies should be considered only as qualitative.

In addition to monomer droplet nucleation, a significant fraction of latex particles are generated by homogeneous nucleation for the lauryl methacrylate containing miniemulsion polymerization system, which exhibits a strong Ostwald ripening effect. Oligomeric radicals generated in the continuous aqueous phase become insoluble when a critical chain length is reached. This water-insoluble free radical may thus coil up and form a particle nucleus (~10^0 nm in diameter). Subsequently, stable primary particles (~10^1 nm in diameter) are produced by the limited flocculation of unstable particle nuclei and adsorption of surfactant molecules on their surfaces. The surfactant species required to stabilize these primary particles may come from those dissolved in the water phase, those released from the shrinking monomer droplet surfaces due to the Ostwald ripening, or even from those adsorbed on the monomer droplet and latex particle surfaces. Homogeneous nucleation becomes less important when the level of lauryl methacrylate is increased (i.e., the Ostwald ripening effect is reduced). On the other hand, particle nucleation in the continuous aqueous phase is greatly suppressed for the polymerization system using the more hydrophobic stearyl methacrylate as the costabilizer. This is simply because the initial monomer droplet size obtained from the stearyl methacrylate-containing polymerization system is smaller in comparison with the lauryl methacrylate counterpart. Thus, the total monomer droplet surface area available for capturing the incoming oligomeric radicals (i.e., the probability for monomer droplet nucleation to take place) is larger for the polymerization system using stearyl methacrylate as the costabilizer. In addition, the larger monomer droplet surface area associated with the stearyl methacrylate-containing polymerization system requires more surfactant for adequate colloidal stability, and the amount of surfactant expelled from the shrinking droplet surfaces due to the Ostwald ripening effect is lower because this reaction system does not show significant diffusional degradation of the droplets. Under these circumstances, the level of free surfactant molecules is not high enough to stabilize most of the particle nuclei generated in the continuous aqueous phase. Such conditions will then promote the absorption of water-borne particle embryos by the latex particles or monomer droplets upon mutual collision induced by the applied shear force. These experimental data indicate that the water solubility of costabilizer has a significant influence on the particle nucleation process and, consequently, the polymerization kinetics.

Increasing the persulfate initiator concentration in miniemulsion polymerization promotes the formation of particle nuclei in the continuous aqueous phase (homogeneous nucleation) [36, 43]. Both the number of latex particles originating from monomer droplet nucleation and the number of water-borne particles increase with increasing the surfactant concentration. It is noteworthy that particle nuclei formed via the micellar nucleation mechanism as expected may become important if the surfactant concentration is well above its critical micelle concentration [44, 45].

Another interesting feature about miniemulsion polymerization is that two kinds of latex particles are obtained when the mixed mode of particle nucleation [i.e., monomer droplet nucleation, homogeneous nucleation, and micellar nucleation (if present)] is operative in the particle formation stage. The population of latex particles arising from monomer droplet nucleation or micellar nucleation contains costabilizer, whereas there is no costabilizer inside the latex particles generated by homogeneous nucleation. Furthermore, these two types of colloidal particles may even show very different surface properties, provided that the costabilizer exhibits some surface activity. For example, cetyl alcohol interacts with sodium dodecyl sulfate strongly and forms intermolecular complexes within the monomer droplet surface layer. Therefore, latex particles generated via the monomer droplet nucleation mechanism are coated with a closely packed structure of surfactant and cosurfactant [5]. In addition, such a closely packed structure at the oil–water interface may act as a barrier to the entry of free radicals and thereby lead to slow monomer droplet nucleation [4, 5, 46, 47]. Formation of particle nuclei takes place up to 40–60% monomer conversion [47]. The probability for the formation of particle nuclei in the continuous aqueous phase will then increase accordingly. As a result of the quite long particle nucleation period associated with the styrene miniemulsion polymerization system using cetyl alcohol as the cosurfactant, the resultant latex product shows a particle size distribution that is broader than that of the conventional emulsion polymerization counterpart.

Long particle nucleation processes (up to 80–90% monomer conversion) were also observed in the styrene miniemulsion polymerization system

Table 5.1. Dependence of Number of Latex Particles per Unit Volume of Water on Surfactant and Initiator Concentrations ([Surfactant]$^{\alpha}$; [Initiator]$^{\beta}$) in Some Representative Miniemulsion Polymerizations

Monomer	Costabilizer	Surfactant	Initiator	α	β	Reference
St	CA	SDS	KPS		0.37	5
St	CA	SDS	AMBN		0.21	5
St	CA	SDS	KPS		0.31	47
St	CA + polymer	SDS	KPS		0	47
St	HD	SDS	KPS		0.11	48
St	HD + polymer	SDS	KPS		0	48
MMA	HD	SDS	KPS	0.77	0.11	50
MMA	PMMA	SDS	KPS		0	26
AN	HD	SDS	AMBN	1.4		51
VAc/BA	HD	SHS	KPS	0.25	0.80	9
VAc/MA	HD	Aerosol MA	KPS		0.63	52

St, styrene; MMA, methyl methacrylate; AN, acrylonitrile; BA, *n*-butyl acrylate; MA, methyl acrylate; VAc, vinyl acetate; CA, cetyl alcohol; HD, hexadecane; PMMA, polymethyl methacrylate; SDS, sodium dodecyl sulfate; SHS, sodium hexyldecyl sulfate; KPS, potassium persulfate; AMBN, 2,2′-azobis-(2-methylbutyronitrile).

containing hexadecane or hexadecane plus a small quantity of preformed polystyrene [48] and in the miniemulsion copolymerization systems of vinyl acetate–*n*-butyl acrylate and vinyl acetate–dioctyl maleate [49]. It is also interesting to note that the number of latex particles per unit volume of water obtained from the hexadecane-containing miniemulsion polymerization system is larger than that obtained from the cetyl alcohol-containing counterpart [53].

5.4.3 Effect of Functional Monomers and Initiators on Particle Nucleation

The presence of hydrophilic, functional monomers such as 2-hydroxyethyl methacrylate, acrylic acid, and methacrylic acid can play an important role in the free radical polymerization taking place in the continuous aqueous phase. Chern and Sheu [54] studied the effect of using a small quantity of 2-hydroxyethyl methacrylate in the styrene miniemulsion polymerization system with sodium dodecyl sulfate and long chain alkyl (lauryl and stearyl) methacrylates as the surfactant and costabilizers, respectively. Both the populations of latex particles originating from monomer droplet nucleation and homogeneous nucleation in the miniemulsion polymerization system in the presence of 2-hydroxyethyl methacrylate become larger as compared to those in the absence of 2-hydroxyethyl methacrylate. Nevertheless, the fraction of latex particles stemming from monomer droplet nucleation decreases with increasing level of 2-hydroxyethyl methacrylate. This is because incorporation of a small amount of 2-hydroxyethyl methacrylate that may exhibit some surface activity into the polymerization system makes the homogenized monomer droplet surfaces more hydrophilic; hence, less surfactant molecules can be adsorbed onto the droplet surfaces. Furthermore, 2-hydroxyethyl methacrylate will enhance the activity of free radical polymerization in the continuous aqueous phase as well. Under these circumstances, at constant surfactant concentration, more oil–water interfacial area can be generated during the homogenization process. As a consequence, a larger population of monomer droplets and stronger homogeneous nucleation can be achieved in the 2-hydroxyethyl methacrylate-containing miniemulsion polymerization system.

When a small amount of water-soluble acrylic acid is used as a functional comonomer in the styrene miniemulsion polymerization, a mixed mode of particle nucleation (monomer droplet nucleation and homogeneous nucleation) is operative [55]. Sodium dodecyl sulfate and alkyl methacrylate (lauryl methacrylate or stearyl methacrylate) were used as the surfactant and costabilizer, respectively, and sodium persulfate was used to initiate the free radical polymerization. It is interesting to note that formation of particle embryos in the continuous aqueous phase becomes less prominent for polymerizations containing acrylic acid in comparison with the polymerizations in the absence of carboxylic monomers. In addition, the extent of homogeneous nucleation decreases with increasing concentration of acrylic acid. It was postulated that incorporation of a small amount of acrylic acid into the reaction mixture results in relatively hydrophilic oligomeric radicals, and this makes the formation of the water-borne particle nuclei more difficult. Furthermore, the efficiency of capturing the incoming oligomeric radicals by the carboxylated latex particles is greatly reduced due to the electrostatic repulsion force between the negatively charged polymer chain segments extended from the particle surfaces and the negatively charged oligomeric radicals ($-SO_4^-$ and $-COO^-$). The probability of nucleation in the submicron monomer droplets is enhanced accordingly. Polymerizations using the less hydrophilic methacrylic acid as the comonomer show an intermediate behavior in the polymerization mechanisms and kinetics.

The influence of the type of initiators (sodium persulfate versus 2,2′-azobisisobutyronitrile) on the particle nucleation mechanisms and kinetics involved in the styrene miniemulsion polymerizations has also been studied [39]. As expected, the oil-soluble 2,2′-azobisisobutyronitrile promotes nucleation in the homogenized monomer droplets. On the other hand, formation of particle nuclei in the continuous aqueous phase becomes more important when water-soluble sodium persulfate is used. This result is consistent with the vinyl chloride work of Saethre et al. [44]. The number of polyvinyl chloride latex particles generated by mechanisms other than monomer droplet nucleation increases with increasing solubility of the initiator in water. In addition, the degree of homogeneous nucleation for miniemulsion polymerizations initiated by hydrogen peroxide is lower than the counterpart initiated by the anionic persulfate initiator.

5.4.4 Polymerization Kinetics

A typical styrene miniemulsion polymerization process using cetyl alcohol or hexadecane as the costabilizer does not show a constant reaction rate period and it can be divided into four major regions based on the polymerization rate versus monomer conversion curve [11, 47], as shown schematically in Figure 5.4b. For comparison, the polymerization rate versus conversion profiles for conventional emulsion polymerization (Figure 5.4a) and microemulsion polymerization (Figure 5.4c) are also included in this figure. First, the rate of miniemulsion polymerization increases rapidly to a primary maximum and then decreases with increasing monomer conversion. This is followed by the increase of polymerization rate to a secondary maximum. After the secondary maximum is achieved, the rate of polymerization then decreases rapidly toward the end of polymerization [47].

The first maximal polymerization rate is attributed to the continuous formation of latex particles (i.e., reaction loci). Formation of latex particles originating from the submicron monomer droplets primarily occurs in Interval I, and the number of reaction loci and the polymerization rate increase with increasing monomer conversion. According to the Smith–Ewart Case 2 kinetics [56], both the number of latex particles per unit volume of water and the concentration of monomer in the particles contribute to the changed rate of polymerization with monomer conversion. The larger the population of latex particles or the concentration of monomer in the particles, the faster the polymerization rate. Therefore, the primary maximal polymerization rate does not necessarily correspond to the end of particle nucleation. Particle nuclei may be generated continuously in Interval II, but this effect may be outweighed by the decreasing polymerization rate due to the reduced concentration of monomer in the latex particles. This unique feature is illustrated by the experimental results of Miller et al. [47]. They showed that the styrene miniemulsion polymerization using cetyl alcohol as the costabilizer exhibits a long and slow particle nucleation. The length of the particle nucleation process is a function of the initiator concentration and the monomer conversion at which particle nucleation ceases is in the range of 40–60%. Monomer droplet nucleation, micellar nucleation for the polymerization system in the presence of micelles, and homogeneous nucleation may take place simultaneously and compete with one another. As a consequence, the resultant latex particle size distribution is negatively skewed with a tail corresponding to the population of small particles. Once a significant concentration of latex particles is generated, the particle nucleation process is greatly retarded because most of the oligomeric radicals are captured by the monomer-swollen parti-

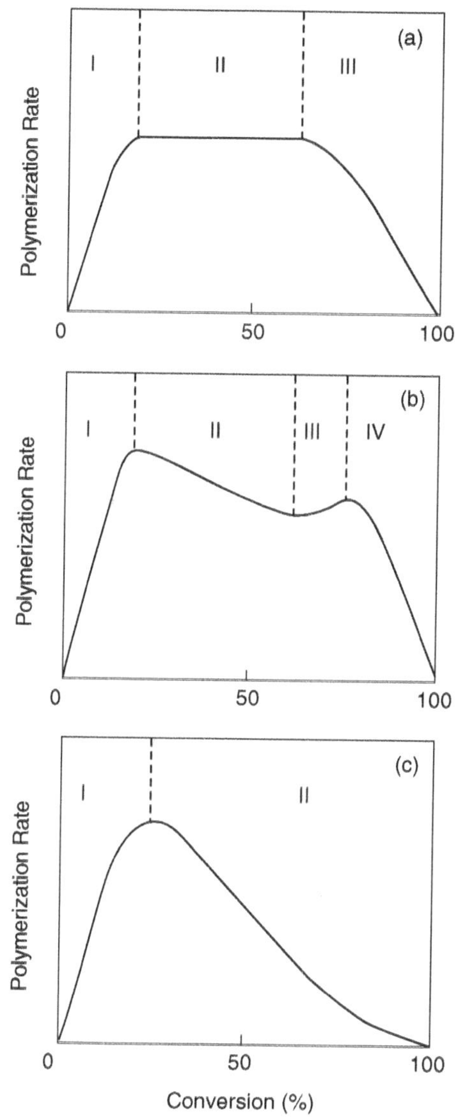

Figure 5.4. A schematic representation of typical polymerization rate as a function of monomer conversion profiles for **(a)** conventional emulsion polymerization (Interval II: Smith–Ewart Case 2 kinetics), **(b)** miniemulsion polymerization, and **(c)** microemulsion polymerization. The distinct intervals of the polymerization processes are also included in these plots.

cles. In addition, the average number of free radicals per particle can be below 0.5, and it increases slowly with increasing monomer conversion in Interval I. This can be attributed to the slow absorption of oligomeric radicals by the monomer droplets.

During Interval II, the concentration of monomer in the reaction loci (latex particles) continues to decrease, and hence the rate of polymerization decreases with increasing monomer conversion. The average number of free radicals per particle is equal to 0.5 during this interval, signifying the "on–off" mechanism upon the entry of an oligomeric radical into the latex particle containing zero (on) or one free radical (off), which is similar to the Smith–Ewart Case 2 kinetics involved in conventional emulsion polymerization [56]. It is also interesting to note that the concentration of monomer and the average number of free radicals in the latex particles originating from monomer droplet nucleation should be different from those in the water-borne particles. This is due to the presence of costabilizer in the nucleated monomer droplets and the quite large size of these highly monomer-swollen particles compared to the water-borne particle nuclei (i.e., **n** may be greater than 0.5).

Beyond Interval II, the second maximal polymerization rate can be attributed to the gel effect. The bimolecular termination reaction becomes diffusion-controlled in the latex particles and the average number of free radicals per particle increases significantly in the latter stage of polymerization, thereby leading to an acceleration of the free radical polymerization. The rate of polymerization then decreases continuously toward the end of polymerization due to the depletion of monomer and/or the diffusion-controlled propagation reaction in the reaction loci.

All the factors discussed here make miniemulsion polymerization kinetics very complicated, as shown by the rather scattered values of the kinetic parameters α, β, γ and δ obtained from the relationships $N_p \sim [S]^\alpha [I]^\beta$ and $R_p \sim [S]^\gamma [I]^\delta$ (e.g., see Table 5.1 in this section and Table 1 in reference 11). The parameter N_p is the number of latex particles per unit volume of water, R_p is the rate of polymerization, and [S] and [I] are the concentrations of surfactant and initiator in water, respectively. As expected, the rate of polymerization in common miniemulsion polymerization systems increases with increasing concentration of surfactant or initiator [11]. As to the influence of the concentration of costabilizer (hexadecane) on the miniemulsion polymerization kinetics, the experimental results reported in the literature are not conclusive [12]. The rate of polymerization may decrease with increasing concentration of hexadecane or this effect may be insignificant in miniemulsion polymerization. These conflicting observations can be attributed to the different concentrations of monomer in the monomer droplets and the varying droplet sizes (or droplet numbers) when the level of hexadecane used in stabilizing the miniemulsion is varied.

In general, the number of latex particles per unit volume of water determined at the end of polymerization and the slope obtained from the least-squares best-fitted linear portion of the monomer conversion versus time curve are taken as the N_p and R_p data in the study of polymerization mechanisms and kinetics. In traditional emulsion polymerization, monomer droplet nucleation, can be neglected and only micellar nucleation, homogeneous nucleation and flocculation of latex particles need to be taken into consideration. It is also quite straightforward to calculate the rate of polymerization using the monomer conversion versus time data obtained from conventional emulsion polymerizations exhibiting a distinct constant reaction rate period. On the other hand, all the particle nucleation mechanisms (monomer droplet nucleation, micellar nucleation, and homogeneous nucleation), coalescence of monomer droplets, and flocculation of latex particles can come into play, and the particle nucleation period is generally quite long in miniemulsion polymerization. Furthermore, the lack of a constant reaction rate interval is often observed in miniemulsion polymerization. These very complex polymerization mechanisms make data analysis work extremely difficult. This is the reason why controversial results are sometimes reported in the literature. More reliable characterization methods are required to study the polymerization mechanisms and kinetics involved in miniemulsion polymerization.

In general, conventional emulsion polymerization is faster in comparison with the miniemulsion polymerization because more latex particles (reaction loci) are nucleated and the rate of polymerization is linearly proportional to the number of latex particles per unit volume of water. Nevertherless, the rate of polymerization per particle is larger for miniemulsion polymerization, as evidenced by the higher concentrations of free radicals and monomer in the latex particles.

The miniemulsion polymerization technique can offer better control over the copolymer composition because incorporation of the constituent monomers into the emulsion polymer product is not governed by the mass transfer process or the water solubility of monomers [9, 57, 58]. However, it can be difficult to produce controlled copolymer compositions because the more reactive monomers will react faster in the monomer droplets than they would in a batch polymerization. It can also be more difficult to produce a latex product with consistent performance properties from batch to batch because of the Ostwald ripening effect experienced in miniemulsion polymerization. The size and number of the homogenized monomer droplets in an inadequately stabilized miniemulsion that exhibits a strong Ostwald ripening effect may vary significantly with the aging time. This may have an influence on the particle nucleation and growth processes in the subsequent free radical polymerization. A fundamental understanding of the colloidal stability of miniemulsions can alleviate such a quality control problem.

5.5 VERSATILITY OF MINIEMULSION POLYMERIZATION

Miniemulsion polymerization is a technique that, in principle, allows any water-insoluble monomer to undergo polymer reactions (not limited to the conventional free radical polymerization) inside the homogenized monomer droplets dispersed in the continuous aqueous phase. Thus, each miniemulsion droplet can be regarded as an ideal submicron scale reactor that is not controlled by the monomer mass transfer process for the synthesis of a variety of

Table 5.2. Representative Examples Reflecting Unlimited Horizon of Miniemulsion Polymerization

Monomer	Costabilizer	Surfactant	Initiator	References
		Catalytic Chain Transfer Reaction		
MMA	HD	SDS	AIBN/KPS	59
		Living Free Radical Polymerization		
St	PS, $C_6F_{13}I$	SDS	ACPA, AIBN	60
St	PS, $C_6F_{13}I$	SDS	ACPA/ $C_6F_{13}I$	61
St	HD	SDBS	KPS/TEMPO	62
St	HD	Dowfax 8390	BPO/TEMPO	63, 64
St	HD, CA	SDS	KPS/$Na_2S_2O_5$, SGI	65
St, MMA	HD, CA	SDS	AIBN or V-40/ dithiobenzoates	66
BMA	HD	Brij 98	V-50/dNbpy/CuBr	67
		Hybrid Miniemulsion Polymers		
Alkyd, MMA, BA, AA	HD	SDS	KPS	68
Sunflower oil, MMA	HD	SDS	Fatty acid hydroperoxide/ Fe^{2+}/SFS/SDTA	69
St, MMA, or BA/HD	HD	SDS	PEGA, KPS, AIBN	70
St/TiO_2	HD, PS	SDS	KPS	71, 72
St/$CaCO_3$	HD	SDS	KPS	73
St/carbon black	Polyester	SDS	KPS, AIBN	74
		Miscellaneous		
Diamine/diepoxide	HD	SDS, Lutensol AT50		75
Diisocyanate/diol	HD	SDS		76

BMA, butyl methacrylate; AA, acrylic acid; AIBN, 2,2′-azobisisobutyronitrile; BPO, benzoyl peroxide.

polymer particles. Some representative examples illustrating the unlimited horizon of the miniemulsion polymerization technique are summarized in Table 5.2. Some of the advanced free radical polymerizations and step polymerizations involved in the miniemulsion polymerization processes will be presented as illustrating examples in the following subsections. Those who are interested in these fascinating areas and others not discussed here are referred to references 12 and 13.

5.5.1 Catalytic Chain Transfer Reaction

Cobalt compounds such as cobaltoxime boron fluoride serve as catalytic chain transfer agents to effectively control the polymer molecular weight in miniemulsion polymerization [59]. It was shown that the solubility of the cobalt catalyst in the reaction system has a significant influence on the polymerization mechanisms. For example, cobaltoxime boron fluoride partitions between the oily phase and the continuous aqueous phase. As a consequence, the free radical polymerization is extremely sensitive to the type of initiators used in miniemulsion polymerization. Poor catalytic activity is greatly retarded for the polymerization system containing oxygen-centered initiator radicals (e.g., those originating from the persulfate initiator), probably due to the poisoning and/or deactivation of the catalyst. On the other hand, tetraphenylcobaltoxime boron fluoride is water-insoluble, and thus its catalytic activity is independent of the type of initiators used to initiate the methyl methacrylate miniemulsion polymerization. This is simply because the catalyst, tetraphenylcobaltoxime boron fluoride, cannot be directly in contact with the initiator radicals. This example demonstrates the general feature of the miniemulsion polymerization system that the discrete submicron monomer droplets effectively provide the catalyst molecules with an isolated hydrophobic environment against the encounter with initiator radicals present in the continuous aqueous phase. This then allows miniemulsion polymerization to proceed without the significant loss of the catalyst activity.

It should be noted that the cobalt catalyst cannot be used to effectively regulate the polymer molecular weight in conventional emulsion polymerization. The cobalt catalyst molecules are primarily present in the emulsified monomer droplets initially. A small proportion of the catalyst may also reside in the monomer-swollen micelles (if present). Transport of the cobalt catalyst molecules from the monomer droplets to the growing latex particles stemming from micellar nucleation or homogeneous nucleation is prohibited during polymerization. Thus, the probability for the chain transfer of a polymeric radical to the cobalt catalyst is greatly reduced.

5.5.2 Living Free Radical Polymerization

Living free radical polymerization is a unique technique used to prepare a variety of polymers with well-controlled molecular structures such as polymers with narrow molecular weight distribution, multiblock copolymers, and star polymers, which cannot be achieved by conventional free radical polymerization [77, 78]. This polymerization technique was originally investigated in homogeneous bulk and solution polymerization systems, and it has been successfully applied to heterogeneous miniemulsion polymerization system in the last decade [62, 63].

For example, controlled free radical polymerization of styrene based on a degenerative transfer process with iodine exchange was carried out in oil-in-water miniemulsions [60, 79]. The synthesis of block copolymers comprising polystyrene and poly(*n*-butyl acrylate) chain segments based on this approach was also demonstrated [61].

The stable free radical polymerization technique is characterized by the growing polymer chains that are reversibly capped by a stable free radical [e.g., 2,2–tetramethyl-1-piperidynyloxy nitroxide (TEMPO)]. For example, stable polystyrene dispersions were prepared by the stable free radical polymerization of styrene conducted in miniemulsion polymerization at 135 °C [62]. Sodium dodecylbenzene sulfonate, hexadecane, and potassium persulfate/TEMPO were used as the surfactant, costabilizer, and initiator system, respectively. Prodpran et al. [63] studied the styrene miniemulsion polymerization stabilized by Dowfax 8390 and hexadecane and initiated by benzoyl peroxide at 125 °C. A molar ratio of TEMPO to benzoyl peroxide equal to 3 to 1 resulted in polystyrene with the lowest polydispersity index (1.3) of polymer molecular weight distribution.

Living free radical polymerizations were also carried out in miniemulsion systems via the reversible addition-fragmentation chain transfer mechanism [66]. The colloidal stability of miniemulsions is the key issue, and nonionic surfactants result in the best results. The polydispersity index of molecular weight distribution for the resultant miniemulsion polymer is generally smaller than 1.2.

Reversible atom transfer free radical polymerization of *n*-butyl acrylate was conducted in miniemulsion systems using the water-soluble initiator 2,2′-azobis(2-amidinopropane) dihydrochloride (V-50) and the hydrophobic ligand 4,4′-di(5-nonyl)-4,4′-bipyridine to form a complex with the copper ions [67, 80]. The resultant Cu(II) complex has a relatively large solubility in the continuous aqueous phase, but this should not impair its capability of controlling the free radical polymerization. This is because the rapid transport of the Cu(II) complex between the dispersed organic phase and the continuous aqueous phase assures an adequate concentration of the free radical deactivator. As a consequence, the controlled free radical polymerization within the homogenized monomer droplets can be achieved.

5.5.3 Step Polymerization

The first step reaction example involves the preparation of water-based epoxy resins via ring-opening polymerizations of different epoxides with various diamines, dithiols or diols residing in the miniemulsion droplets at 60 °C [75]. The basic requirement for successful miniemulsion polymerization is that both reactive components exhibit relatively low solubility in the continuous aqueous phase. The diepoxide Epikote E828, triepoxide Decanol EX-314, and tetraepoxide EX-411 are potential candidates for this purpose. Furthermore, incorporation of conventional costabilizers such as hexadecane into the miniemulsion polymerization system does not improve the stability of these miniemulsions because these extremely hydrophobic epoxides also act as reactive costabilizers in the retardation of the Ostwald ripening effect. Suitable diamines include Jeffamine D2000 (an NH_2-terminated polypropylene oxide with an average molecular weight of 2032 g mol^{-1}, 4,4′-diaminobibenzyl, 1,12-diaminododecane, and 4,4′-diaminodicyclohexylmethane. Jeffamins D400, with the lowest molecular weight in this series of diamine compounds, is too hydrophilic to be used to synthesize epoxy resins in miniemulsion systems. In addition to diamine compounds, both 1,6-hexanedithiol and bisphenol A were evaluated in this work. Based on gel permeation chromatography, the average molecular weights of the resultant miniemulsion polymers comprising epoxides and 1,6-hexanedithiol (or bisphenol A) are about 2×10^4 g mol^{-1}, with a polydispersity index of ~2. This implies that the intimate contact of the oil droplet surfaces with water molecules does not show any appreciable negative effect on the ring-opening polymerization taking place in the miniemulsion droplets.

Water-based polyurethane products can also be prepared by homogenizing a mixture of diisocyanates and diols in an aqueous surfactant solution, followed by heating the resultant miniemulsion to the prescribed step polymerization temperature [76]. Satisfactory miniemulsion polyurethanes are obtained only when the following requirements are fulfilled.

(a) The solubility of diisocyanates and diols in water should be very low.
(b) The reaction between the isocyanate group and the hydroxyl group should be much slower than the time required for the homogenization step to form the miniemulsion.
(c) The side reaction between the isocyanate group and water to form a urea group should be slower than the urethane formation in the miniemulsion droplets.

The reaction between the isocyanate groups near the droplet surface layer and water molecules are thought to generate more hydrophobic urea groups that form a passivated surface layer to retard further reaction of the isocyanate groups with water inside the droplets.

The molar ratio of isocyanate/hydroxyl is set at 1/1. In the absence of undesired side reactions, this molar ratio will lead to polyurethane chains with very high molecular weight in the discrete polymer particles dispersed in the continuous aqueous phase. A significant advantage to the end-users in handling this type of polyurethane is that these latex products generally show excellent rheological properties as compared to those prepared by bulk or solution polymerization techniques. Experimental results show the successful formation of polyurethane. Furthermore, the side reaction between the isocyanate groups near the droplet surface layer and water molecules does occur during the miniemulsion polymerization. Nevertheless, this side reaction is only of secondary importance in the preparation of polyurethane using the miniemulsion polymerization technique.

REFERENCES

1. J. Ugelstad, M. S. El-Aasser, and J. W. Vanderhoff, *J. Polym. Sci., Polym. Lett. Ed.* **11**, 503 (1973).
2. J. Ugelstad, F. K. Hansen, and S. Lange, *Dielektr. Makromol. Chem.* **175**, 507 (1974).
3. D. P. Durbin, M. S. El-Aasser, G. W. Poehlein, and J. W. Vanderhoff, *J. Appl. Polym. Sci.* **24**. 703 (1979).
4. B. J. Chamberlain, D. H. Napper, and R. G. Gilbert, *J. Chem. Soc. Faraday Trans. 1* **78**, 591 (1982).
5. Y. T. Choi, M. S. El-Aasser, E. D. Sudol, and J. W. Vanderhoff, J. *Polym. Sci. Polym. Chem. Ed.* **23**, 2973 (1985).
6. J. Delgado, M. S. El-Aasser, and J. W. Vanderhoff, *J. Polym. Sci. Polym. Chem. Ed.* **24**, 861 (1986).
7. J. Delgado, M. S. El-Aasser, C. A. Silebi, J. W. Vanderhoff, and J. Guillot, *J. Polym. Sci. Polym. Phys. Ed.* **26**, 1495 (1988).
8. J. Delgado, M. S. El-Aasser, C. A. Silebi, and J. W. Vanderhoff, *J. Polym. Sci. Polym. Chem. Ed.* **27**, 193 (1989).
9. J. Delgado, M. S. El-Aasser, C. A. Silebi, and J. W. Vanderhoff, *J. Polym. Sci. Polym. Chem. Ed.* **28**, 777 (1990).
10. P. L. Tang, E. D. Sudol, C. A. Silebi, and M. S. El-Aasser, *J. Appl. Polym. Sci.* **43**, 1059 (1991).
11. I. Capek and C. S. Chern, *Adv. Polym. Sci.* **155**, 101 (2001).
12. M. Antonietti and K. Landfester, *Prog. Polym. Sci.* **27**, 689 (2002).
13. J. M. Asua, *Prog. Polym. Sci.* **27**, 1283 (2002).
14. G. W. Poehlein, private communication, 2007.
15. A. S. Kabalnov and E. D. Shchukin, *Adv. Colloid Interface Sci.* **38**, 69 (1992).
16. I. M. Lifshitz and V. V. Slezov, *J. Phys. Chem. Solids.* **19**, 35 (1961).
17. C. Wagner, *Ber. Bunsenges. Phys. Chem.* **65**, 581 (1961).
18. P. Taylor, *Adv. Colloid Interface Sci.* **75**, 107 (1998).
19. A. S. Kabalnov, K. N. Makarov, A. V. Pertzov, and E. D. Shchukin, *J. Colloid Interface Sci.* **138**, 98 (1990).
20. W. I. Higuchi and J. Misra, *J. Pharm. Sci.* **51**, 459 (1962).
21. S. Wang and F. J. Schork, *J. Appl. Polym. Sci.* **54**, 2157 (1994).
22. C. M. Miller, P. J. Blythe, E. D. Sudol, C. A. Silebi, and M. S. El-Aasser, *J. Polym. Sci. Polym. Chem. Ed.* **32**, 2365 (1994).
23. C. M. Miller, E. D. Sudol, C. A. Silebi, and M. S. El-Aasser, *Macromolecules* **28**, 2754 (1995).

24. C. M. Miller, E. D. Sudol, C. A. Silebi, and M. S. El-Aasser, *Macromolecules* **28**, 2765 (1995).

25. C. M. Miller, E. D. Sudol, C. A. Silebi, and M. S. El-Aasser, *Macromolecules* **28**, 2772 (1995).

26. J. Reimers and F. J. Schork, *J. Appl. Polym. Sci.* **59**, 1833 (1996).

27. J. L. Reimers and F. J. Schork, *J. Appl. Polym. Sci.* **60**, 251 (1996).

28. J. A. Alduncin, J. Forcada, and J. M. Asua, *Macromolecules* **27**, 2256 (1994).

29. J. L. Reimers and F. J. Schork, *Ind. Eng. Chem. Res.* **36**, 1085 (1997).

30. D. Mouran, J. Reimers, and F. J. Schork, *J. Polym. Sci. Polym. Chem. Ed.* **34**, 1073 (1996).

31. S. Wang, G. W. Poehlein, and F. J. Schork, *J. Polym. Sci. Polym. Chem. Ed.* **35**, 595 (1997).

32. C. S. Chern, T. J. Chen, and Y. C. Liou, *Polymer* **39**, 3767 (1998).

33. C. S. Chern and T. J. Chen, *Colloid Polym. Sci.* **275**, 546 (1997).

34. C. S. Chern and T. J. Chen, *Colloid Polym. Sci.* **275**, 1060 (1997).

35. C. S. Chern and T. J. Chen, *Colloid Surface A* **138**, 65 (1998).

36. C. S. Chern, Y. C. Liou, and T. J. Chen, *Macromol. Chem. Phys.* **199**, 1315 (1998).

37. C. S. Chern and Y. C. Liou, *Macromol. Chem. Phys.* **199**, 2051 (1998).

38. C. S. Chern and Y. C. Liou, *Polymer* **40**, 3763 (1999).

39. C. S. Chern and Y. C. Liou, *J. Polym. Sci. Polym. Chem. Ed.* **37**, 2537 (1999).

40. K. Landfester, N. Bechthold, S. Foster, and M. Antonietti, *Macromol. Rapid Commun.* **20**, 81 (1999).

41. H. C. Chang, Y. Y. Lin, C. S. Chern, and S. Y. Lin, *Langmuir* **14**, 6632 (1998).

42. N. Bechthold and K. Landfester, *Macromolecules* **33**, 4682 (2000).

43. H. Huang, H. Zhang, J. Li, S. Cheng, F. Hu, and B. Tan, *J. Appl. Polym. Sci.* **68**, 2029 (1998).

44. B. Saethre, P. C. Mork, and J. Ugelstad, *J. Polym. Sci. Polym. Chem. Ed.* **33**, 2951 (1995).

45. M. S. Lim and H. Chen, *J. Polym. Sci. Polym. Chem. Ed.* **38**, 1818 (2000).

46. F. K. Hansen and J. Ugelstad, *J. Polym. Sci. Polym. Chem. Ed.* **17**, 3069 (1979).

47. C. M. Miller, E. D. Sudol, C. A. Silebi, and M. S. El-Aasser, *J. Polym. Sci. Polym. Chem. Ed.* **33**, 1391 (1995).

48. P. J. Blythe, B. R. Morrison, K. A. Mathauer, E. D. Sudol, and M. S. El-Aasser, *Langmuir* **16**, 898 (2000).

49. X. Q. Wu and F. J. Schork, *Ind. Eng. Chem. Res.* **39**, 2855 (2000).

50. K. Fontenot and F. J. Schork, *J. Appl. Polym. Sci.* **49**, 633 (1993).

51. K. Landfester and M. Antonietti, *Macromol. Rapid Commun.* **21**, 820 (2000).

52. J. Delgado and M. S. El-Aasser, *Macromol. Chem. Macromol. Symp.* **31**, 63 (1990).

53. J. Ugelstad, P. C. Mork, K. H. Kaggerud, T. Ellingsen, and A. Berge, *Adv. Colloid Interface Sci.* **13**, 101 (1980).

54. C. S. Chern and J. C. Sheu, *J. Polym. Sci. Polym. Chem. Ed.* **38**, 3188 (2000).
55. C. S. Chern and J. C. Sheu, *Polymer* **42**, 2349 (2001).
56. W. V. Smith and R. H. Ewart, *J. Chem. Phys.* **16**, 592 (1948).
57. V. S. Rodriguez, Interparticle Monomer Transport in Miniemulsion Copolymerization, Ph.D. Dissertation, Lehigh University, Bethlehem, PA, 1988.
58. J. Reimers and F. J. Schork, *Polym. React. Eng.* **4**, 135 (1996).
59. D. Kukulj, T. P. Davis, and R. G. Gilbert, *Macromolecules* **30**, 7661 (1997).
60. M. Lansalot, C. Farcet, B. Charleux, and J. P. Vairon, *Macromolecules* **32**, 7354 (1999).
61. C. Farcet, M. Lansalot, R. Pirri, J. P. Vairon, and B. Charleux, *Macromol. Rapid Commun.* **21**, 921 (2000).
62. P. J. MacLeod, R. Barber, P. G. Odell, B. Keoshkerian, and M. K. Georges, *Macromol. Symp.* **155**, 31 (2000).
63. T. Prodpran, V. L. Dimonie, E. D. Sudol, and M. S. El-Aasser, *Macromol. Symp.* **155**, 1 (2000).
64. G. Pan, E. D. Sudol, V. L. Dimonie, and M. S. El-Aasser, *Macromolecules* **34**, 481 (2001).
65. C. Farcet, M. Lansalot, B. Charleux, R. Pirri, and J. P. Vairon, *Macromolecules* **33**, 8559 (2000).
66. H. de Brouwer, J. G. Tsavalas, F. J. Schork, and M. J. Monteiro, *Macromolecules* **33**, 9239 (2000).
67. K. Matyaszewski, J. Qiu, N. V. Tsarevsky, and B. Charleux, *J. Polym. Sci. Polym. Chem. Ed.* **38**, 4724 (2000).
68. S. T. Wang, F. J. Schork, G. W. Poehlein, and J. W. Gooch, *J. Appl. Polym. Sci.* **60**, 2069 (1996).
69. E. M. S. van Hamersveld, J. van Es, and F. P. Cuperus, *Colloid Surface A: Physicochem. Eng. Asp.* **153**, 285 (1999).
70. F. Tiarks, K. Landfester, and M. Antonietti, *Langmuir* **17**, 908 (2001).
71. B. Erdem, E. D. Sudol, V. L. Dimonie, and M. S. El-Aasser, *J. Polym. Sci. Polym. Chem. Ed.* **38**, 4431 (2000).
72. B. Erdem, E. D. Sudol, V. L. Dimonie, and M. S. El-Aasser, *Macromol. Symp.* **155**, 181 (2000).
73. N. Bechthold, F. Tiarks, M. Willert, K. Landfester, and M. Antonietti, *Macromol. Symp.* **151**, 549 (2000).
74. F. Tiarks, K. Landfester, and M. Antonietti, *Macromol. Chem. Phys.* **202**, 51 (2001).
75. K. Landfester, F. Tiarks, H. P. Hentze, and M. Antonietti, *Macromol. Chem. Phys.* **201**, 1 (2000).
76. F. Tiarks, K. Landfester, and M. Antonietti, *J. Polym. Sci. Polym. Chem. Ed.* **39**, 2520 (2001).
77. K. F. O'Driscoll and S. Russo, Eds. *International Symposium On Free Radical Polymerization, Macromol. Symp.* **111**, 1–328 (1996).

78. K. Matyjaszewski, in *Controlled/Living Radical Polymerization. Progress in ATRP, NMP, and RAFT*, ACS Symposium Series, Vol. 768, Washington, D.C. 2000, pp. 2–26.

79. A. Butte, G. Storti, and M. Morbidelli, *Macromolecules* **33**, 3485 (2000).

80. J. Qiu, T. Pintauer, S. G. Gaynor, K. Matyaszewski, B. Charleux, and J. P. Vairon, *Macromolecules* **33**, 7310 (2000).

6

MICROEMULSION POLYMERIZATION

An oil-in-water (O/W) or water-in-oil (W/O) microemulsion product consists of fine oil (or water) droplets (~10^0–10^1 nm in diameter) dispersed in the continuous aqueous (or oily) phase with the aid of surfactant and/or cosurfactant (e.g., sodium dodecyl sulfate and *n*-pentanol for O/W microemulsion) [1]. The pioneering work on microemulsion was established by Schulman and coworkers [2, 3]. Extensive studies on this unique colloidal system were carried out after the 1974 oil crisis. This then led to great interest in free radical polymerization in vinyl-monomer-containing microemulsions. The polymer colloid products generally exhibit small latex particles and very-high-polymer molecular weight and find a wide range of potential applications. This chapter presents a brief review of the formation and microstructure of the thermodynamically stable microemulsion. The structural characteristic of the initial microemulsion is expected to have a significant influence on the subsequent free radical polymerization inside the very fine monomer droplets. Polymerization mechanisms and kinetics in microemulsion polymerization are the primary focuses of this chapter. Another goal is to introduce the concept of using this kind of self-organized surfactant solution (microemulsion) as a template to synthesize organic polymers.

6.1 INTRODUCTION

In the preparation of an O/W microemulsion, incorporation of amphipathic cosurfactant into the adsorbed layer of anionic surfactant around the oil

Principles and Applications of Emulsion Polymerization, by Chorng-Shyan Chern

droplet greatly reduces the electrostatic repulsion force between two adjacent surfactant molecules, lowers the oil–water interfacial tension (i.e., the change of Gibbs free energy after the formation of microemulsion) to be close to zero, and decreases the persistence length of the interfacial layer (i.e., enhances the flexibility of the interfacial membrane). All these synergistic factors promote the spontaneous formation of transparent one-phase microemulsions exhibiting excellent fluidity. Unlike the classical emulsion, the transparent or translucent reaction system comprising microemulsion droplets is thermodynamically stable in nature, and these tiny droplets exhibit an extremely large oil–water interfacial area ($\sim 10^5\,m^2\,dm^{-3}$).

The concept of free radical polymerization in microemulsion droplets was established by the work of Schauber in 1979 [4]. Microemulsion polymerization involves the propagation reaction of free radicals with vinyl monomer molecules in very fine oil (or water) droplets dispersed in the continuous aqueous (or oily) phase. Relatively stable polymer particles ($\sim 10^1$ nm in diameter) consisting of only a few polymer chains per particle are produced, and therefore the resultant polymer molecular weight is very high (in the range 10^6–$10^7\,g\,mol^{-1}$). This cannot be achieved readily by conventional emulsion polymerization (Chapters 3 and 4) or miniemulsion polymerization (Chapter 5). In addition, the particle nucleation and growth mechanisms and kinetics associated with microemulsion polymerization are quite different from those of emulsion and miniemulsion polymerization systems. Research interests in microemulsion polymerization techniques have grown rapidly since the 1980s because of their potential applications in the preparation of fine latex particles, ultrahigh-molecular-weight water-soluble polymers (flocculants), novel porous materials, polymeric supports for binding metal ions, conducting polymers, colloidal particles containing various functional groups for the biomedical field, and transparent colloidal systems for photochemical and other chemical reactions. A number of representative review articles dealing with microemulsion polymerization are available [5–12].

6.2 FORMATION AND MICROSTRUCTURE OF MICROEMULSIONS

6.2.1 Formation of Microemulsions

Considering a colloidal system at constant temperature, volume, and composition, the change of Helmholtz free energy (dF) for any process undergoing the expansion of the oil–water interfacial area ($dA > 0$) can be expressed as

$$dF/dA = \sigma - W_{des} \tag{6.1}$$

where σ is the oil–water interfacial tension and W_{des} is the work of desorption of surfactant per unit interfacial area. The term W_{des} is the resultant of various components such as changes in entropy and surface charge density and molecular interactions between constituents of the interfacial film. When W_{des} is sufficiently large and σ is extremely small (about 10^{-3} mN m^{-1} or lower), dF/dA becomes negative and, therefore, dF is smaller than zero ($\because dA > 0$) after the formation of the microemulsion. Under these circumstances, a spontaneous reduction in the average oil droplet size down to 10^{0}–10^{1} nm is achieved, thereby leading to a thermodynamically stable colloidal system (microemulsion). For example, a gradual transition from the milky O/W emulsion to the transparent microemulsion and then to the lamellar gel phase can be observed visually when n-pentanol (approximately 0.4 g each time) is added successively to the mixture comprising 100 g of water, 11 g of sodium dodecyl sulfate, and 5.92 g of styrene, with mild mixing at room temperature. It should be noted that cosurfactant (n-pentanol in this case) is not a must in the preparation of microemulsion.

Figure 6.1 shows a schematic representation of the pseudo three-component phase diagram of a typical (surfactant/cosurfactant)–oil–water system. Depending on formulas, fine oil droplets dispersed in the continuous aqueous phase [O/W (or direct) microemulsion] or water droplets in the continuous oily phase [W/O (or inverse) microemulsion] are obtained. Furthermore, the intermediate region between the O/W microemulsion phase and the W/O microemulsion phase is characterized by a bicontinuous microstructure in which the aqueous and oily microdomains are interconnected with each other [13, 14]. The presence of such a middle phase in the colloidal system was verified by literature data [15]. It was shown that the oil–water interfacial layer in the bicontinuous microstructure has a zero mean curvature (i.e., it is flat on the average), and this sponge-like microstructure is completely disor-

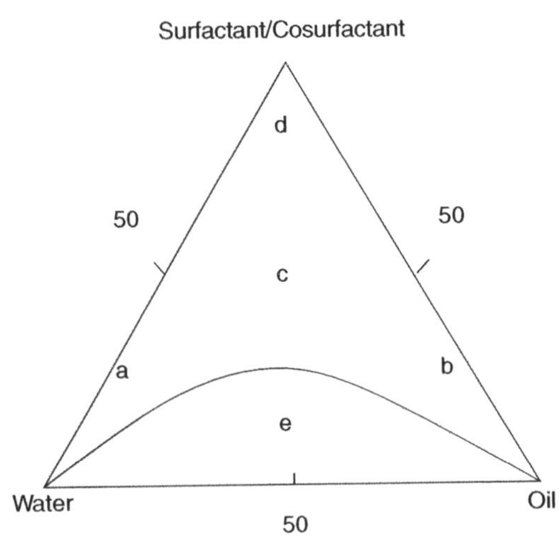

Figure 6.1. A schematic representation of the pseudo three-phase diagram of (surfactant/cosurfactant)–oil–water systems: **(a)** oil-in-water microemulsion, **(b)** water-in-oil microemulsion, **(c)** bicontinuous structure, **(d)** lamellar structure, and **(e)** conventional two-phase emulsion.

dered. Such a structural aspect of microemulsions allows the vinyl monomer to be incorporated into different phases in the polymerization system [e.g., the dispersed phase (microemulsion droplets), the continuous phase, or the bicontinuous phase]. This then makes it practical to design and prepare versatile microemulsion polymers for a variety of applications. Information on the initial condition of the reaction system (e.g., the type of microemulsion, the dimension of microemulsion droplets, the physical properties of the oil–water interfacial film, the distance between two droplets, and the viscosity of microemulsion) is required to gain a better understanding of the subsequent polymerization mechanisms and kinetics.

6.2.2 Factors that Govern Microemulsion Structures

The type of microemulsions (O/W or W/O) is primarily governed by the spontaneous or preferred curvature of the oil–water interfacial layer (i.e., the surfactant layer). By convention, the preferred curvature of the interfacial layer is positive for O/W microemulsions, whereas it is negative for W/O microemulsions. For dilute microemulsion systems, the preferred curvature of the interfacial layer can be manipulated by varying the ratio of surfactant to cosurfactant. This compositional parameter allows swelling of the oil (or water) droplets until a maximal level of swelling is reached, at which the radius of the droplets is close to the reciprocal of the curvature of the interfacial layer. However, for microemulsion systems with relatively large volume fractions of the dispersed phase, it is the attractive interparticle interactions that govern the degree of swelling of the droplets.

A transition from an O/W microemulsion to a W/O microemulsion and vice versa can be induced by continuously changing the ratio of oil to water. The type of microstructures in the phase inversion domain is primarily controlled by the bending constant, which is a characteristic of the elasticity of the surfactant layer [16]. If the magnitude of the bending constant is only on the order of kT, where k is the Boltzmann constant and T the absolute temperature, the persistence length of the oil–water interfacial film (i.e., the distance over which the interfacial film is locally flat) is microscopically small. In this case, the interfacial film is flexible and it is easily deformed under thermal fluctuations. Thus, the phase inversion process takes place via the bicontinuous structure that consists of randomly interconnected oil and water microdomains [13, 14]. This colloidal system is characterized by a mean curvature of about zero (or a flat interfacial film on the average) [17] and a maximal solubilization capacity. On the other hand, the persistence length of the oil–water interfacial film is large and the interfacial film is flat over a macroscopic distance when the bending constant is much greater than the thermal energy, kT. Under these circumstance, the phase inversion transition process then occurs through a lamellar phase [16].

In addition to simple O/W and W/O microemulsions, a variety of phases can coexist in equilibrium in a single microemulsion [18]. For example, Winsor

I is a globular O/W microemulsion in equilibrium with an excess of oil, Winsor II is a globular W/O microemulsion in equilibrium with an excess of water, and Winsor III is a microemulsion with a bicontinuous microstructure that is in equilibrium with both the oily and aqueous phases.

A unique characteristic of microemulsions is the transient behavior of the aggregates. Surfactant molecules and other constituent species are constantly exchanged among the oil (or water) droplets. These droplets collide with one another and then form aggregates with an average lifetime on the order of microseconds or even longer for the colloidal system in the presence of attractive interactions [19, 20]. Thus, one may envision that free radical polymerization taking place in these dynamic microemulsion droplets will be accompanied by structural changes [7]. It should be noted that spontaneous formation of thermodynamically stable microemulsions generally requires a small amount (only a few percent) of monomer along with a very large amount (>10%) of surfactant due to the extremely large oil–water interfacial area that needs to be stabilized. This may severely limit the potential applications of microemulsion polymerization because water-based polymer products with high solids contents and low levels of surfactant are desirable for most industrial applications.

6.3 O/W MICROEMULSION POLYMERIZATION

6.3.1 General Features

Significant research efforts have been devoted to the free radical polymerization of relatively hydrophobic monomers such as styrene, methyl methacrylate, and *n*-butyl acrylate in O/W microemulsions. The anionic surfactant sodium dodecyl sulfate in combination with a short-chain alcohol (e.g., *n*-pentanol) as the cosurfactant is the most popular stabilization package used in common microemulsion polymerization systems. However, as more polymer forms with the progress of polymerization, the increase of free energy as a result of the conformational limitation and/or incompatibility between polymer and cosurfactant results in the colloidal instability or turbidity of microemulsion polymer [21]. When a cationic surfactant, dodecyltrimethylammonium bromide, is used alone to stabilize the styrene microemulsion polymerization, eventually no cosurfactant is required to form satisfactory microemulsion products [22]. It should be noted that the amount of monomer that can be solubilized within microemulsion droplets is generally quite low, whereas the levels of surfactant and cosurfactant required to stabilize the colloidal system are very high in O/W microemulsion products. For example, the weight percentage of styrene does not exceed a few percent, which is much lower than that of surfactant/cosurfactant (about 16% total) used in typical microemulsion polymerizations.

Just like in other heterogeneous polymerization techniques, the presence of a continuous aqueous or oily phase allows satisfactory control of the reaction temperature and the product is an ultrahigh-molecular-weight polymer dispersion that exhibits excellent fluidity. Furthermore, unlike conventional emulsion and miniemulsion polymerization systems, at constant pressure, the initial state of a microemulsion immediately before the start of polymerization is dependent only on composition and temperature because microemulsions are thermodynamically stable. This characteristic allows intrinsic control and reproducibility of the performance properties of microemulsion polymers. The resultant latex particles are typically characterized by a dimension of about 5–100 nm, a rather narrow particle size distribution, and a very small number of polymer chains per particle.

Significant efforts have been devoted to microemulsion polymerization mechanisms and kinetics since the 1980s. In comparison with other self-organized microemulsion structures (e.g., the bicontinuous structure), a much simpler colloidal structure comprising the discrete oil (or water) droplets dispersed in the continuous aqueous (or oily) phase simplifies the investigation of the interactions among monomer, polymer, surfactant, and water and, therefore, can help isolate the particle nucleation and growth mechanisms and the monomer transport phenomenon [11]. Thus, studies of the O/W (or W/O) microemulsion polymerization systems can provide a fundamental understanding of the complex polymerization mechanisms and kinetics in self-organized reaction media.

6.3.2 Polymerization Mechanisms and Kinetics

The O/W microemulsion polymerization system, (sodium dodecyl sulfate-*n*-pentanol)/styrene/water, has been investigated extensively during the past 20 years. Representative studies carried out by Guo et al. [23–26] are chosen for discussion hereinafter. Sodium persulfate and 2,2′-azobis(2-methyl butyronitrile) were used as the water-soluble and oil-soluble initiators, respectively, in their experiments. The rate of polymerization first increases to a maximum with the progress of the reaction [Interval I, ranging from 0 to 20–25% monomer conversion for microemulsion polymerizations initiated by sodium persulfate and ranging from 0 to 10–15% conversion for polymerizations initiated by 2,2′-azobis(2-methyl butyronitrile)]. The polymerization rate then decreases toward the end of the reaction (Interval II). A schematic representation of the rate of polymerization versus monomer conversion profile for the O/W microemulsion polymerization is shown in Figure 5.4c. The constant rate of polymerization period, often observed in conventional emulsion polymerization (Figure 5.4a), is absent from microemulsion polymerization. Furthermore, the gel effect (i.e., the rapidly increased monomer conversion with time owing to the greatly retarded bimolecular termination reaction) is not experienced in the latter stage of microemulsion polymerization. This is simply because the resultant latex particles (20–30 nm in diameter) are not large enough to accommodate more than one free radical therein.

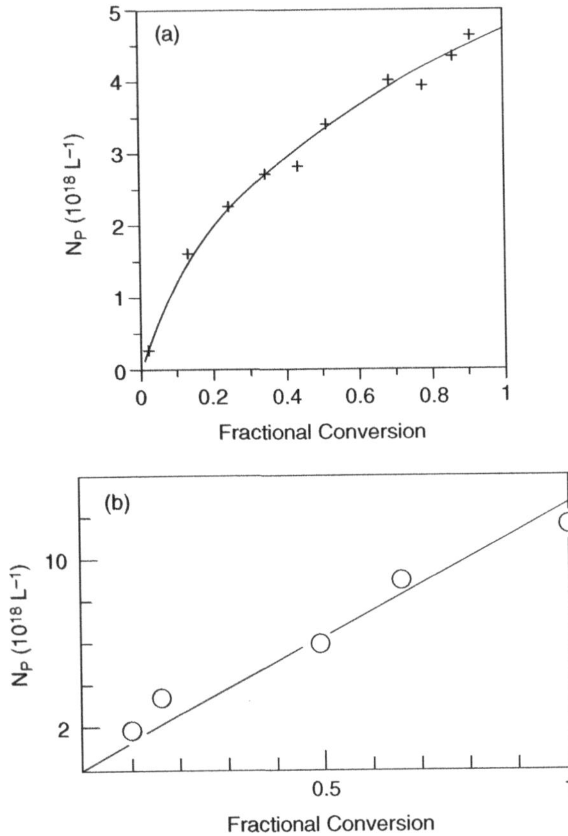

Figure 6.2. Number of latex particles per unit volume of water as a function of monomer conversion for **(a)** O/W microemulsion polymerization of styrene [24] and **(b)** W/O microemulsion polymerization of acrylamide [61].

Guo et al. [23–26] also showed that nucleation of latex particles occurs throughout the microemulsion polymerization, as shown in Figure 6.2a. Microemulsion droplets (i.e., monomer-swollen micelles with an average diameter of 4 nm) are primary loci for the nucleation of latex particles. However, nucleation of particle nuclei in the continuous aqueous phase (homogeneous nucleation [27–29]) and limited flocculation of these particle nuclei cannot be completely ruled out. The molecular weight of microemulsion polystyrene obtained was very large (1×10^6–2×10^6 g mol^{-1}) due to the predominant chain transfer of polymeric radicals to monomer molecules inside the growing latex particles. The resultant monomeric radicals may desorb out of the very small latex particles, thereby leading to a value of the average number of free radicals per particle smaller than 0.5. These monomeric radicals in the continuous aqueous phase may also be reabsorbed by the latex particles. A mechanistic model was developed to simulate the reaction kinetics of the styrene microemulsion polymerization system. The entry rate coefficient of free radicals into the microemulsion droplets estimated by this model was 7×10^5 cm^3 mol^{-1} s^{-1}, which is several orders of magnitude smaller than that of free radicals into the monomer-swollen polymer particles. This is most likely due to the condensed interfacial layer of sodium dodecyl sulfate and *n*-pentanol on the microemulsion droplet surface, which acts as a physical barrier to the incoming free radicals. As a result, only a small fraction (about 10^{-3}) of the microemulsion droplets initially present in the polymerization system can be successfully transformed into latex particles. The resultant latex particles are thus much larger than the microemulsion droplets initially present in the polymerization system. Therefore, microemulsion polymerization can be regarded as a reconstructive template synthesis [30]. In this series of studies conducted by Guo et al. [23–26], a satisfactory agreement between the experimental data and model predictions was achieved. However, the extremely small entry rate coefficient of free radicals into the microemulsion droplets was considered to be physically unreasonable because this would lead to the greatly enhanced bimolecular termination reaction rate in Guo's model [31].

Morgan et al. [31] derived the following compact equation for predicting microemulsion polymerizaqtion kinetics:

$$\ln(1-X) = -fk_d[\mathrm{I}]_0\, k_p[\mathrm{M}]_{d,0}\, t^2/[\mathrm{M}]_0 \tag{6.2}$$

where X is the monomer conversion, f is the initiator efficiency factor, k_d is the initiator decomposition rate constant, k_p is the propagation reaction rate constant, t is the reaction time, and $[M]_{d,0}$ and $[M]_0$ are the initial concentrations of monomer in the microemulsion droplets and in the polymerization system, respectively. This kinetic model was developed based on the assumptions that (a) all the free radicals generated in the continuous aqueous phase can enter the microemulsion droplets (i.e., particle nucleation loci) and then initiate the free radical polymerization therein, (b) the bimolecular termination reaction taking place in the aqueous phase is negligible, (c) the absorption of free radicals by the latex particles is insignificant, and (d) the growing polymeric radicals in a latex particle is terminated primarily by the monomer chain transfer reaction. According to Eq. (6.2), plotting ln $(1 - X)$ versus t^2 data should result in a straight line with a slope of $(-fk_d[I]_0k_p[M]_{d,0}/[M]_0)$. This simple model predicts the kinetic behavior of the *n*-hexyl methacrylate microemulsion polymerization reasonably well. In fact, it is the only existing kinetic model that can predict the course of microemulsion polymerization, as shown in Figure 6.3 [32]. The general validity of this kinetic model was assessed by several sets of experimental data obtained from styrene microemulsion polymerization, and poor performance of Eq. (6.2) in the high monomer conversion region was often observed [32–35]. Such a discrepancy between the experimental data and the model predictions can be attributed to the

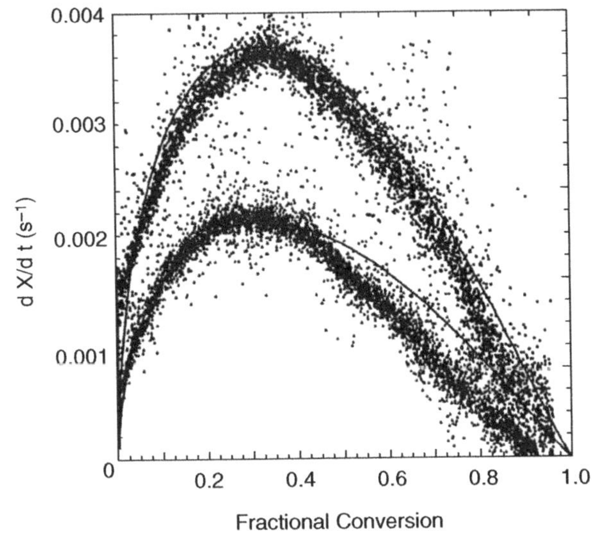

Figure 6.3. Rate of polymerization (*dX/dt*) versus monomer conversion profiles for the microemulsion polymerization of *n*-hexyl methacrylate stabilized by a cationic surfactant, dodecyltrimethylammonium bromide [32]. The two curves represent the microemulsion polymerizations with different concentrations of initiator [initiator/monomer = 0.045 wt% (top) and 0.015 wt% (bottom)]. The discrete points represent the experimental data, and the solid lines the model predictions according to Eq. (6.2).

bimolecular termination and diffusion-controlled propagation reactions in the latex particles [32].

A number of investigations deal with the O/W microemulsion polymerizations of styrene or methyl methacrylate [22, 24, 25, 36–44]. These studies showed that the solubility of monomer in water has a significant influence on the polymerization mechanisms and kinetics. It was postulated that nucleation in the microemulsion droplets predominates in the polymerization of styrene, which has a very limited water solubility (0.031%). Homogeneous nucleation is considered negligible due to the effective capture of free radicals generated in the continuous aqueous phase by the very large population of microemulsion droplets. On the other hand, polymerization of the relatively hydrophilic methyl methacrylate (water solubility = 1.56%) in O/W microemulsions initiated by a water-soluble initiator may result in competitive particle nucleation mechanism (microemulsion droplet nucleation versus homogeneous nucleation) because the residence time for the oligomeric radicals generated in the continuous aqueous phase is long enough to allow them to grow to a critical chain length and then form particle nuclei. Bleger et al. [44] proposed that homogeneous nucleation occurs early in the methyl methacrylate microemulsion polymerization, in which the rate of polymerization is very slow. This is followed by the predominant nucleation in the microemulsion droplets and the rate of polymerization becomes much faster.

The creative technique of pyrene fluorescene intensity measurements was proposed to investigate the particle nucleation mechanisms involved in the O/W microemulsion polymerization [45]. The experimental data show that microemulsion droplets are the major particle nucleation loci for the polymerization system with the more hydrophobic styrene as the monomer. This is followed by the flocculation of the latex particles with the remaining droplets. In contrast, the free radical polymerization taking place initially in the continuous aqueous phase (homogeneous nucleation) plays an important role in methyl methacrylate microemulsion polymerization. The computer simulation work of Mendizabal et al. [46] also led to the conclusion that the extent of homogeneous nucleation increases with increasing the solubility of monomer in water.

The feasibility of using a water-insoluble dye to study the particle nucleation mechanisms in the styrene microemulsion polymerization was evaluated [35]. The parameters chosen for this work include the type of cosurfactants (*n*-butanol, *n*-pentanol, and *n*-hexanol), the concentration of the persulfate initiator, and the type of initiator (the oil-soluble 2,2′-azobisisobutyronitrile versus water-soluble sodium persulfate). The rationale behind this approach is that most of the dye molecules are solubilized within the microemulsion droplets immediately before the start of polymerization. Thus, dye molecules can be incorporated into the latex particles only when free radicals originating from sodium persulfate are absorbed by these dye containing droplets. This is followed by the free radical polymerization inside the microemulsion droplets, which then successfully converts these droplets into particle nuclei. Furthermore, primary particles generated in the continuous aqueous phase (homogeneous nucleation) should not contain any dye species because the transport of dye molecules from the microemulsion droplets or latex particles originating from nucleation in the droplets, across the continuous aqueous phase, and then into the latex particles originating from homogeneous nucleation is prohibited due to the extremely low water solubility of dye. Thus, the weight percentage of dye ultimately incorporated into the final latex particles serves as an indicator for the extent of nucleation in the microemulsion droplets. It was shown that the dye content in the resultant latex particles increases with increasing the alkyl chain length of alcohols. Based on these experimental data, the importance of homogeneous nucleation in decreasing order is *n*-butanol > *n*-pentanol > *n*-hexanol. Guo et al. [23] illustrated that the number of latex particles nucleated per unit volume of water increases with increasing the persulfate initiator concentration. This trend can be attributed to the enhanced flux of free radicals into the microemulsion droplets. It was suggested in reference 35 that the increase of the sodium persulfate concentration not only promotes the capture of free radicals by the microemulsion droplets but also increases the probability for oligomeric radicals to precipitate out of the continuous aqueous phase and then form particle nuclei. The latter effect may override the former, thereby leading to the decreased dye content in the resultant latex particles with the sodium persulfate concentration.

Chern and Wu [35] also studied the influence of the oil-soluble initiator, 2,2′-azobisisobutyronitrile (AIBN), on the particle nucleation and growth mechanisms involved in styrene microemulsion polymerization. Considering the limiting case that the particle nucleation process is controlled predominately by the formation of AIBN radicals in the monomer phase, the dye content in the resultant latex particles should be much larger compared to the styrene microemulsion polymerization initiated by sodium dodecyl sulfate. However, this was not the case because the bimolecular termination reaction between two neighboring AIBN radicals produced as pairs in the very small microemulsion droplets is significant. It was then postulated that the particle nucleation process involves (a) the desorption of one free radical out of the microemulsion droplet containing two AIBN radicals, followed by the propagation reaction between the remaining AIBN radical and monomer molecules, (b) the absorption of one free radical by the droplet already containing two AIBN radicals, followed by the bimolecular termination reaction and the subsequent propagation reaction between the survivor and monomer molecules, or (c) the entry of one free radical into the droplet containing no free radicals, followed by the propagation reaction between this free radical and monomer molecules therein. For the styrene microemulsion polymerization initiated by AIBN, the particle nucleation process is characterized by a very low efficiency for the initiation reaction. Under these circumstances, a much higher AIBN concentration is required to give a dye content in the resultant latex particles that is comparable to that of the styrene microemulsion polymerization initiated by sodium persulfate.

Recently, Chern and Tang [47] adopted the water-insoluble dye technique to study microemulsion polymerization mechanisms and kinetics. It was shown that both the number of latex particles per unit volume of water and the rate of polymerization increase with increasing the initiator concentration for the styrene microemulsion polymerization system stabilized by sodium dodecyl sulfate/*n*-pentanol and initiated by sodium persulfate. Least-squares best-fitting experimental data with a modified Morgan–Nomura model (Eq. (6.2)) [31, 33] led to the conclusion that the limited flocculation of latex particles plays an important role in the particle growth process. The weight percentage of dye incorporated into the latex particles increases continuously with the progress of polymerization. This implies that particle nucleation takes place throughout the polymerization. Furthermore, at constant monomer conversion, the weight percentage of dye incorporated into the latex particles decreases with increasing the initiator concentration, which is attributed to the increased probability of particle nucleation in the continuous aqueous phase with the higher initiator concentration. The authors also illustrated that the relatively low initiation efficiency of the oil-soluble initiator, AIBN, makes the styrene microemulsion polymerization system stabilized by sodium dodecyl sulfate/*n*-pentanol display a quite different particle nucleation mechanism from that initiated by the water-soluble sodium persulfate [48]. Formation of particle nuclei in the continuous aqueous phase can be suppressed to some extent and microemulsion droplet nucleation predominates in the styrene polymerization initiated by AIBN. The effects of monomers with different water solubility (i.e., the relatively hydrophobic styrene versus the relatively hydrophilic methyl methacrylate) on the microemulsion polymerizations initiated by sodium persulfate are dramatic. Homogeneous nucleation plays an important role in the particle formation process, and a mixed mode of particle nucleation (microemulsion droplet nucleation and homogeneous nucleation) is operative in the methyl methacrylate microemulsion polymerization. The very high level of surfactant often used to prepare microemulsions and the desorbed surfactant molecules from the microemulsion droplet surfaces promote the formation of particle nuclei in the continuous aqueous phase. The methyl methacrylate microemulsion polymerization experiences stronger flocculation of the latex particles in comparison with the styrene counterpart. This is closely related to the difference in the surface coverage with sodium dodecyl sulfate and *n*-pentanol between the polymethyl methacrylate and polystyrene particles during polymerization. The polymethyl methacrylate particles with a lower surface charge density show a greater tendency to flocculate with one another in order to reduce the total particle surface area and enhance the colloidal stability.

Kaler and co-workers [31, 32, 49–53] contributed a series of excellent papers dealing with the effects of a number of kinetic parameters (e.g., the partitioning of monomer, bimolecular termination reaction, and diffusion-controlled propagation reaction) on the reaction kinetics, latex particle size distribution, and polymer molecular weight distribution in various O/W microemulsion polymerization systems. Analytical expressions based on a mechanistic model were developed to predict the latex particle size distribution and polymer molecular weight distribution [49]. These computer simulation results were then verified by experimental data obtained from the quasielastic light scattering and gel permeation chromatography measurements [49] and the on-line small-angle neutron scattering technique [50]. It is also interesting to note that the small-angle neutron scattering data indicate that the *n*-hexyl methacrylate monomer does not swell polymer particles during the microemulsion polymerization [50]. Mixed cationic surfactants of dodecyltrimethylammonium bromide and didodecyldimethylammonium bromide were used to stabilize the microemulsion polymerization system. A model based on the latex particle comprising a polymer core surrounded by a monomer-rich shell is consistent with the polymerization kinetic data and small-angle neutron scattering measurements provided that the concentration of monomer in the shell of the latex particles is equal to that in the core of the *n*-hexyl methacrylate monomer-swollen micelles.

The small-angle neutron scattering technique was further adopted to study the partitioning of monomer between the microemulsion droplets and latex particles for a variety of monomers (styrene, *n*-butyl methacrylate, *t*-butyl methacrylate, and *n*-hexyl methacrylate) [51]. The monomer partitioning behavior determines the concentration of monomer in the growing latex particles and plays an important role in the particle nucleation and growth processes [49, 50, 54]. It was found that, during microemulsion polymerization, the partitioning of monomer is strongly dependent on the composition of the microemulsion, especially on the distance to the phase boundary in the pseudo three-phase diagram of the heterogeneous (surfactant/cosurfactant)–oil–water system. For example, the partitioning of monomer is linear in nature and the concentration of monomer in the latex particles is quite low if the initial microemulsion composition is far away from the phase boundary. In contrast, the partitioning of monomer is essentially nonlinear and the concentration of monomer in the latex particles is much higher if the initial microemulsion composition is quite close to the phase boundary. The experimental data can be interpreted reasonably well by modeling the monomer partitioning behavior as a competition between the Flory–Huggins bulk polymer free energy and the Helfrich curvature elastic energy of the surfactant monolayer. The failure of Eq. (6.2) can be attributed to a combined effect of the nonlinear monomer partitioning behavior, bimolecular termination reaction in the latex particles, and, in some cases, the diffusion-limited propagation reaction [52]. It is also very interesting to note that the weight-average molecular weight of the resultant polystyrene, close to $1.5 \times 10^7\,g\,mol^{-1}$, about one order of magnitude greater than that controlled by the chain transfer reaction of polymeric radical to monomer ($\sim 2 \times 10^6\,g\,mol^{-1}$), was measured by gel permeation chromatography coupled with a multiangle laser light scattering detector and a differential refractive index detector [53]. This can be attributed to the diffusion-controlled desorption of monomeric radicals out of the latex particles in combination with the chain transfer reaction of a polymeric radical to polymer.

In addition to common short-chain alcohols [35, 55], other compounds such as diethylene glycol monoalkyl ether were evaluated as potential cosurfactants in the styrene microemulsion polymerization with sodium dodecyl sulfate as the surfactant [56]. Candau [7] pointed out that the main difficulty in achieving microemulsions with higher levels of monomer lies in retaining the optical transparency and colloidal stability of the microemulsions upon free radical polymerization. In addition to entropic factors, the compatibility between polymer and cosurfactant also contributes to the destabilization of microemulsions during polymerization [21]. This is especially true when styrene undergoes free radical polymerization within the O/W microemulsion droplets containing a short-chain alcohol because the cosurfactant is a poor solvent for polystyrene. The presence of short-chain alcohols may alter the partitioning of monomer in microemulsion systems [57]. Furthermore, short-chain alcohols act as chain transfer agents that lower the polymer molecular weight during microemulsion polymerization [6]. By contrast, the acrylic monomer *n*-butyl acrylate can partition into the oil–water interfacial layer to some extent. This promotes the formation of *n*-butyl acrylate microemulsions in the absence of a typical cosurfactant such as *n*-pentanol [58].

It is interesting to note that the type of polymerization techniques (conventional emulsion polymerization versus microemulsion polymerization) can have a significant influence on the chain transfer reactions and, consequently, polymer properties [59]. For example, in the microemulsion polymerization of vinyl acetate, the chain transfer reaction of a polymeric radical to monomer is the predominant mechanism that terminates the free radical reactivity. Thus, the resultant polyvinyl acetate exhibits a lower degree of branching than that produced by the conventional emulsion polymerization process, in which the chain transfer reaction of a polymeric radical to polymer is significant.

Although the results obtained from these studies have made significant progress in this field, further research is still required to gain a better understanding of microemulsion polymerization mechanisms and kinetics.

6.4 W/O MICROEMULSION POLYMERIZATION

The most widely investigated hydrophilic monomer in the W/O microemulsion polymerizations is acrylamide. In this case, acrylamide is usually dissolved in water [50/50 (w/w)] immediately before the preparation of microemulsion because acrylamide is in powder form at room temperature. Thus, the resultant polyacrylamide particles are swollen with water and dispersed in the continuous organic medium. These polymer products are widely used for the flocculation of colloidal dispersions, sewage treatment, coatings, adhesives, and enhanced oil recovery. Just like the hydrophobic monomer *n*-butyl acrylate in the formation of an O/W microemulsion [58], acrylamide also tends to migrate toward the water–oil interfacial layer and it acts as a reactive cosurfactant in the formation of a W/O microemulsion using toluene as the continuous phase [60, 61]. Sodium bis(2-ethylhexyl)sulfosuccinate (AOT) is the anionic surfactant of choice. The free radical polymerization can be initiated by thermal initiators, photochemical initiators, or γ-radiolysis. A number of representative studies of acrylamide within the microemulsion droplets are included in references 62–80. The W/O microemulsion polymerization systems also suffer from the drawback that a very high ratio of surfactant to monomer (2.5–25) is generally required to stabilize the colloidal system. The basic formulas, characteristics, and kinetic parameters of some representative W/O microemulsion polymerization systems can be found in two excellent review articles [6, 7].

Candau et al. [64, 65] proposed a mechanistic model for particle nucleation and growth in the W/O microemulsion polymerization system comprising sodium bis(2-ethylhexyl)sulfosuccinate, acrylamide, water, and toluene. A schematic model for the nucleation and growth of latex particles in acrylamide microemulsion polymerization stabilized by sodium bis(2-ethylhexyl)sulfosuccinate is shown in Figure 6.4 [81]. Based on elastic and quasi-elastic light scattering, viscometry, and ultracentrifugation experiments, they observed that the diameter of the resultant latex particles (in the range of 20–40 nm) is

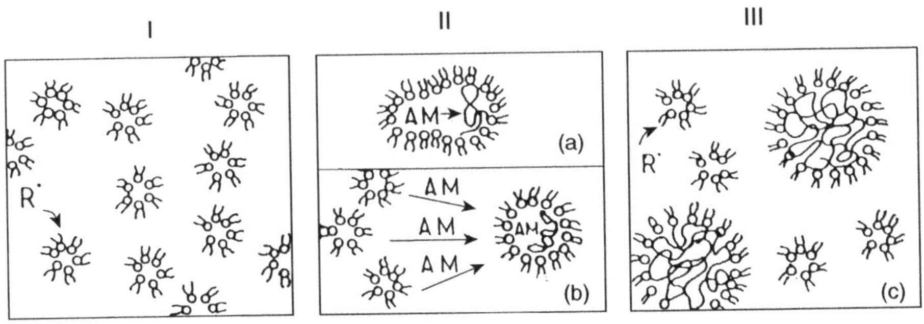

Figure 6.4. A schematic model for the nucleation and growth of latex particles in the acrylamide microemulsion polymerization stabilized by sodium bis(2-ethylhexyl)sulfosuccinate. **(I)** The initial condition of the polymerization system consists of a very large population of the acrylamide/water-swollen micelles (~6 nm in diameter) dispersed in the continuous oily phase. Nucleation of particle nuclei occurs when free radicals are absorbed by the microemulsion droplets. **(II)** Growth of latex particles are achieved by **(a)** collision and then coalescence between two particles and **(b)** diffusion of monomer molecules from the microemulsion droplets through the continuous oily phase and then into the particles. **(c)** The polymerization system comprises water-swollen polyacrylamide particles (~40 nm in diameter) and acrylamide/water-swollen micelles (~3 nm in diameter) dispersed in the continuous oily phase at the end of polymerization [81].

much larger than that of initial microemulsion droplets (~6 nm); that is, the final number of latex particles per unit volume of water is approximately two to three orders of magnitude smaller than the number of microemulsion droplets initially present in the polymerization system. Furthermore, on the average, only one polymer chain per particle is achieved in the acrylamide microemulsion polymerization. Based on these experimental results, it was postulated that latex particles are nucleated continuously throughout the polymerization, as illustrated in Figure 6.2b. This is not the case for the conventional emulsion polymerization in which the nucleation of latex particles stops at about 10–20% monomer conversion [27–29].

The growth of latex particles is attributed to the very limited fraction of the initial microemulsion droplets that can be successfully converted into particle nuclei. The remaining microemulsion droplets only serve as reservoirs to supply the growing particle nuclei with monomer and surfactant during polymerization. Transport of monomer molecules from microemulsion droplets to latex particles can be achieved either by molecular diffusion through the continuous oily phase or by the mutual collision between one droplet and one particle. Monomer-swollen micelles (i.e., microemulsion droplets) are present throughout the polymerization because of the very high level of surfactant used to stabilize the colloidal system. The probability for primary radicals generated in the continuous organic phase to enter these monomer-swollen micelles with an extremely large monomer/water–oil interfacial area is much higher in comparison with latex particles. Thus, nucleation of latex particles is carried into the latter stage of polymerization, and each particle, on the average, is capable of capturing only one free radical during its life time.

The above particle nucleation and growth mechanisms were verified experimentally [61]. It was shown that the number density of latex particles increases linearly with increasing monomer conversion. On the other hand, the latex particle size remains relatively constant with the progress of polymerization. Such reaction mechanisms with some minor modifications may also be adequate for the qualitative description of the nucleation and growth of latex particles in the O/W microemulsion polymerization.

For the W/O microemulsion polymerization of acrylamide stabilized by sodium bis(2-ethylhexyl)sulfosuccinate and initiated by 2,2′-azobisisobutyronitrile, the initiation reactions take place predominantly in the acrylamide/water-toluene interfacial layer, in which the encounter of initiator radicals with monomer molecules is facilitated [70–74]. On the other hand, as would be expected, free radical polymerization is initiated primarily within the acrylamide/water cores of the microemulsion droplets when the water-soluble persulfate initiator is used. The technique of steady-state fluorescence of indolic probes quenched by acrylamide and selectively located in different phases (the continuous toluene phase, the acrylamide/water–oil interface and the acrylamide/water phase) of the W/O microemulsion system stabilized by sodium bis(2-ethylhexyl)sulfosuccinate was adopted to study the consumption of monomer during polymerization [79]. The experimental results show that acrylamide is consumed evenly from all parts of the microemulsion polymerization system, regardless of the initial microemulsion composition and the nature of initiator.

6.5 POLYMERIZATION IN CONTINUOUS OR BICONTINUOUS PHASES OF MICROEMULSIONS

In an attempt to prepare hydrophobic polymers designed to encapsulate water-soluble materials, the feasibility of using W/O microemulsion polymerization systems was evaluated in the 1980s. For example, the influence of free radical polymerization within the continuous organic phase (styrene and acrylic monomers such as methyl methacrylate and methyl acrylate) on the regions of stable W/O microemulsion systems was investigated [21, 81–84]. It was observed that formation of polymer results in serious phase separation during polymerization. Therefore, a much smaller region of stable microemulsions can be obtained in comparison with the counterpart in the absence of polymer. These results imply that a recipe that is adequate to the preparation of satisfactory microemulsion does not guarantee the successful transformation of this microemulsion into a unique microemulsion polymer product. The factors responsible for this instability problem include changes in the molecular conformation (the entropic effect) and interactions among constituents (e.g., monomer, polymer and cosurfactant).

Porous polymeric materials can be synthesized by the free radical polymerization of styrene in Winsor I, II, and III microemulsions [85, 86]. It was found that the porosity of polymer achieved by polymerization in the bicontinuous phase was higher than that in the continuous phase of the O/W or W/O microemulsion. This was attributed to the interconnected microdomains in the bicontinuous phase. In fact, the microstructure of polymer achieved is closely related to the nature of microemulsion [87–91]. For example, porous polymer with a closed-cell structure (i.e., the discrete water pores are distributed throughout the polymer matrix) forms when the polymerization of the W/O microemulsion takes place. In contrast, polymerization in the bicontinuous phase results in porous polymer with an open-cell structure, in which water channels are interconnected throughout the polymer matrix. These results strongly indicate that the morphology of the resultant porous polymer retains the initial structure of microemulsion to some extent. Nevertheless, the dimension of the resultant porous microdomains ($\sim 10^3$ nm) is much larger than that of the initial structure of microemulsion ($<10^2$ nm) as a consequence of phase separation during the progress of polymerization. A breakthrough showed that it is possible to produce polymer with a pore size of 50–70 nm, which is on the order of that of the precursor microemulsion [92]. It is very interesting to note that polymer nanoparticles with a very narrow particle size distribution can be produced by polymerization in and phase separation from the bicontinuous phase [93]. Hentze and Kaler [11] pointed out that this process might be an interesting alternative to the unstable and turbid W/O latexes obtained from classical inverse emulsion polymerization, because the nanoparticles prepared by microemulsion polymerization are very stable. In general, polymers prepared by polymerization within the bicontinuous phase are characterized by a slightly lower molecular weight as compared to those obtained from polymerization in the discrete fine droplets of microemulsions [94].

The potential applications of such a polymerization technique for preparing novel polymeric materials include microfiltration, separation membranes, polymer blends with a unique microstructural morphology, and porous microcarriers for cultures of living cells and enzymes [7]. Some other interesting ideas about the preparation of novel materials include the conductive composite film [95] and microporous silica gel [96].

REFERENCES

1. L. M. Prince (Ed.), *Microemulsions: Theory and Practice*, Academic Press, New York, 1977.
2. J. H. Schulman and D. P. Riley, *J. Colloid Sci.* **3**, 383 (1948).
3. J. H. Schulman and J. A. Friend, *J. Colloid Sci.* **4**, 497 (1949).
4. C. Schauber, These Docteur-Ingenieur, Universite de Mulhouse, France, 1979.

5. F. Candau, in *Encyclopedia of Polymer Science and Engineering*, H. F. Mark, N. M. Bikales, C. G. Overberger, and G. Mengers (Eds.), Wiley-Interscience, New York, 1987, pp. 718–724.
6. F. Candau, in *Polymerization in Organized Media*, C. M. Paleos (Ed.), Gordon & Breach, Philadelphia, 1992, Chapter 4, pp. 215–282.
7. F. Candau, in *Handbook of Microemulsion Science and Technology*, P. Kumar and K. L. Mittal (Eds.), Marcel Dekker, New York, 1999, pp. 679–712.
8. I. Capek, *Adv. Colloid Interface Sci.* **80**, 85 (1999).
9. I. Capek, *Adv. Colloid Interface Sci.* **82**, 253 (1999).
10. C. S. Chern, in *Encyclopedia of Polymer Science and Technology*, J. I. Kroschwitz (Ed.), John Wiley & Sons, New York, 2002.
11. H. P. Hentze and E. W. Kaler, *Curr. Opin. Colloid Interface Sci.* **8**, 164 (2003).
12. X. J. Xu and L. M. Gan, *Curr. Opin. Colloid Interface Sci.* **10**, 239 (2005).
13. L. E. Scriven, *Nature (London)* **263**, 123 (1976).
14. S. Friberg, I. Lapczynska, and G. Gillberg, *J. Colloid Interface Sci.* **56**, 19 (1976).
15. L. Auvray, J. P. Cotton, R. Ober, and C. Taupin, *J. Phys. Chem.* **88**, 4586 (1984).
16. P. G. de Gennes and C. Taupin, *J. Phys. Chem.* **92**, 2294 (1982).
17. L. Auvray, J. P. Cotton, R. Ober, and C. Taupin, *J. Phys.* **45**, 913 (1984).
18. P. Winsor, *Solvent Properties of Amphiphilic Compounds*, Butterworth, London, 1954.
19. P. Guering, A. M. Cazabat, and M. Paillette, *EuroPhys. Lett.* **2**, 953 (1986).
20. D. Langevin, *Annu. Rev. Phys. Chem.* **43**, 341 (1992).
21. L. M. Gan, C. H. Chew, and S. E. Friberg, *J. Macromol. Sci., Chem. A* **19**, 739 (1983).
22. V. H. Perez-Puna, J. E. Puig, V. M. Castano, B. E. Rodriguez, A. K. Murthy, and E. Kaler, *Langmuir* **6**, 1040 (1990).
23. J. S. Guo, M. S. El-Aasser, and J. W. Vanderhoff, *J. Polym. Sci., Polym. Chem. Ed.* **27**, 691 (1989).
24. J. S. Guo, E. D. Sudol, J. W. Vanderhoff, and M. S. El-Aasser, *J. Polym. Sci., Polym. Chem. Ed.* **30**, 691 (1992).
25. J. S. Guo, E. D. Sudol, J. W. Vanderhoff, and M. S. El-Aasser, *J. Polym. Sci., Polym. Chem. Ed.* **30**, 703 (1992).
26. J. S. Guo, E. D. Sudol, J. W. Vanderhoff, H. J. Yue, and M. S. El-Aasser, *J.Colloid Interface. Sci.* **149**, 184 (1992).
27. R. G. Gilbert, *Emulsion Polymerization: A Mechanistic Approach*, Academic Press, New York, 1995.
28. M. Nomura, H. Tobita, and K. Suzuki, *Adv. Polym. Sci.* **175**, 1 (2005).
29. C. S. Chern, *Prog. Polym. Sci.* **31**, 443 (2006).
30. E. J. Paul and R. K. Prud'homme, in *Reactions and Synthesis in Surfactant Systems*, J. Texter (Ed.), Surfactant Science Series, Vol. **100**, Marcel Dekker, New York, 2001, pp. 525–535.
31. J. D. Morgan, K. M. Lusvardi, and E. W. Kaler, *Macromolecules* **30**, 1897 (1997).
32. R. de Vries, C. C. Co, and E. W. Kaler, *Macromolecules* **34**, 3233 (2001).

33. M. Nomura and K. Suzuki, *Macromol. Chem. Phys.* **198**, 3025 (1997).
34. C. S. Chern and L. J. Wu, *J. Polym. Sci., Polym. Chem. Ed.* **39**, 898 (2001).
35. C. S. Chern and L. J. Wu, *J. Polym. Sci., Polym. Chem. Ed.* **39**, 3199 (2001).
36. L. Feng and K. Y. S. Ng, *Macromolecules* **23**, 1048 (1990).
37. L. Feng and K. Y. S. Ng, *Colloids Surf.* **53**, 349 (1991).
38. J. E. Puig, V. H. Perez-Luna, M. Perez-Gonzales, E. R. Macias, B. E. Rodriguez, and E. W. Kaler, *Colloid Polym. Sci.* **271**, 114 (1993).
39. L. M. Gan, C. H. Chew, K. C. Lee, and S. C. Ng, *Polymer* **34**, 3064 (1993).
40. L. M. Gan, C. H. Chew, S. C. Ng, and S. E. Loh, *Langmuir* **9**, 2799 (1993).
41. L. M. Gan, C. H. Chew, K. C. Lee, and S. C. Ng, *Polymer* **35**, 2659 (1994).
42. L. M. Gan, C. H. Chew, J. H. Lim, K. C. Lee, and L. H. Gan, *Colloid Polym. Sci.* **272**, 1082 (1994).
43. L. A. Rodriguez-Guadarrama, E. Mendizabal, J. E. Puig, and E. W. Kaler, *J. Appl. Polym. Sci.* **48**, 775 (1993).
44. F. Bleger, A. K. Murthy, F. Pla, and E. W. Kaler, *Macromolecules* **27**, 2559 (1994).
45. D. R. Kim and D. H. Napper, *Macromol. Rapid Commun.* **17**, 845 (1996).
46. E. Mendizabal, J. Flores, J. E. Puig, I. Katime, F. Lopez-Serrano, and J. Alvarez, *Macromol. Chem. Phys.* **201**, 1259 (2000).
47. C. S. Chern and H. J. Tang, *Polym. React. Eng.* **11**, 213 (2003).
48. C. S. Chern and H. J. Tang, *J. Appl. Polym. Sci.* **97**, 2005 (2005).
49. J. D. Morgan and E. W. Kaler, *Macromolecules* **31**, 3197 (1998).
50. C. C. Co and E. W. Kaler, *Macromolecules* **31**, 3203 (1998).
51. C. C. Co, R. de Vries, and E. W. Kaler, *Macromolecules* **34**, 3224 (2001).
52. R. de Vries, C. C. Co, and E. W. Kaler, *Macromolecules* **34**, 3233 (2001).
53. C. C. Co, P. Cotts, S. Burauer, R. de Vries, and E. W. Kaler, *Macromolecules* **34**, 3245 (2001).
54. P. G. Sanghvi, N. K. Pokhriyal, and S. Devi, *J. Appl. Polym. Sci.* **84**, 1832 (2002).
55. C. S. Chern and C. W. Liu, *Colloid Polym. Sci.* **278**, 329 (2000).
56. C. S. Chern and C. W. Liu, *Colloid Polym. Sci.* **278**, 821 (2000).
57. J. S. Guo, M. S. El-Aasser, E. D. Sudol, H. J. Yu, and J. W. Vanderhoff, *J. Colloid Interface Sci.* **140**, 175 (1990).
58. I. Capek, V. Juranicova, J. Barton, J. M. Asua, and K. Ito, *Polym. Int.* **43**, 1 (1997).
59. R. G. Lopez, M. E. Trvino, R. D. Peralta, L. C. Cesteros, I. Katime, J. Flores, E. Mendizabal, and J. E. Puig, *Macromolecules* **33**, 2848 (2000).
60. F. Candau, Y. S. Leong, G. Pouyet, and S. J. Candau, *J. Colloid Interface Sci.* **101**, 167 (1984).
61. M. T. Carver, E. Hirsch, J. C. Wittmann, R. M. Fitch, and F. Candau, *J. Phys. Chem.* **93**, 4867 (1989).
62. Y. S. Leong, G. Riess, and F. Candau, *J. Chim. Phys.* **78**, 279 (1981).
63. Y. S. Leong, S. J. Candau, and F. Candau, in *Surfactants in Solution*, Vol. 3, K. L. Mittal and B. Lindman (Eds.), Plenum Press, New York, 1984, pp. 1897–1910.

64. F. Candau, Y. S. Leong, G. Pouyet, and S. J. Candau, *J. Colloid Interface Sci.* **101**, 167 (1984).

65. F. Candau, Y. S. Leong, and R. M. Fitch, *J. Polym. Sci., Polym. Chem. Ed.* **23**, 193 (1985).

66. F. Candau, in *Comprehensive Polymer Science: Chain Polymerization II*, Vol. 4, G. G. Eastmond, A. Ledwith and P. Sigwalt (Eds.), Pergamon, Oxford, 1988, Chapter 13, pp. 225–229.

67. M. T. Carver, U. Dreyer, R. Knoesel, F. Candau, and R. M. Fitch, *J. Polym. Sci., Polym. Chem. Ed.* **27**, 2161 (1989).

68. M. T. Carver, F. Candau, and R. M. Fitch, *J. Polym. Sci., Polym. Chem. Ed.* **27**, 2179 (1989).

69. C. Holtzscherer, S. J. Candau, and F. Candau, in *Surfactants in Solution*, Vol. 6, K. L. Mittal and P. Bothorel (Eds.), Plenum Press, New York, 1986, pp. 1473–1481.

70. V. Vaskova, V. Juranicova, and J. Barton, *Angew. Makromol. Chem.* **191**, 717 (1990).

71. V. Vaskova, V. Juranicova, and J. Barton, *Angew. Makromol. Chem.* **192**, 989 (1991).

72. V. Vaskova, V. Juranicova, and J. Barton, *Makromol. Chem.* **192**, 1339 (1991).

73. J. Barton, *Makromol. Chem., Rapid Commun.* **12**, 675 (1991).

74. V. Vaskova, Z. Hlouskova, J. Barton, and V. Juranicova, *Makromol. Chem.* **193**, 627 (1992).

75. J. Barton, *Polym. Int.* **30**, 151 (1993).

76. J. Barton, J. Tino, Z. Hlouskova, and M. Stillhammerova, *Polym. Int.* **34**, 89 (1994).

77. I. Janigova, K. Csomorova, M. Stillhammerova, and J. Barton, *Makromol. Chem. Phys.* **195**, 3609 (1994).

78. V. Vaskova, M. Stillhammerova, and J. Barton, *Chem. Pap.* **48**, 355 (1994).

79. I. Lacik, J. Barton, and G. G. Warr, *Makromol. Chem. Phys.* **196**, 2223 (1995).

80. C. Daubresse, C. Grandfils, R. Jerome, and P. Teyssie, *J. Colloid Interface Sci.* **168**, 222 (1994).

81. F. Candau, in *An Introduction to Polymer Colloids*, F. Candau and R. Ottewill (Eds.), NATO ASI Ser. CNo. 303, Kluwer, Dordrecht, 1990, Chapter 3, pp. 73–96.

82. J. O. Stoffer and T. Bone, *J. Polym. Sci., Polym. Chem. Ed.* **18**, 2641 (1980).

83. J. O. Stoffer and T. Bone, *J. Dispersion Sci. Technol.* **1**, 37 (1980).

84. J. O. Stoffer and T. Bone, *J. Dispersion Sci. Technol.* **4**, 393 (1980).

85. E. Haque and S. Qutubuddin, *J. Polym. Sci., Polym. Lett. Ed.* **26**, 429 (1988).

86. S. Qutubuddin, E. Haque, W. J. Benton, and E. J. Fendler, in Polymer Association Structures: *Microemulsions and Liquid Crystals*, M. El-Nokaly (Ed.), ACS Symposium Series No. 384, American Chemical Society, Washington, D.C., 1989, pp. 65.

87. M. Sasthav and H. M. Cheung, *Langmuir* **7**, 1378 (1991).

88. W. R. P. Raj, M. Sasthav, and H. M. Cheung, *Langmuir* **7**, 2586 (1991).

89. M. Sasthav, W. R. P. Raj, and H. M. Cheung, *J. Colloid Interface Sci.* **152**, 376 (1992).

90. W. R. P. Raj, M. Sasthav, and H. M. Cheung, *Langmuir* **8**, 1931 (1992).
91. W. R. P. Raj, M. Sasthav, and H. M. Cheung, *J. Appl. Polym. Sci.* **47**, 499 (1993).
92. L. M. Gan, T. D. Li, C. H. Chew, W. K. Teo, and L. H. Gan, *Langmuir* **11**, 3316 (1995).
93. F. Candau, M. Pabon, and J. Y. Anquetil, *Colloids Surface A: Physicochem. Eng. Aspects* **153**, 47 (1999).
94. J. Hao, *J. Polym. Sci., Polym. Chem. Ed.* **39**, 3320 (2001).
95. D. A. Kaplin and S. Qutubuddin, *Synth. Met.* **63**(3), 187 (1994).
96. J. H. Burban, H. E. Mengtao, and E. L. Cussler, *AIChE J.* **41**(1), 159 (1995).

7

SEMIBATCH AND CONTINUOUS EMULSION POLYMERIZATIONS

In addition to batch emulsion polymerization (which is commonly used in the laboratory to study reaction mechanisms) for preliminary development/screening of new latex products and to obtain approximate kinetic data for process development and reactor scale-up, the versatile semibatch and continuous emulsion polymerization processes are widely used for the production of commercial latex products. A major reason that batch reactors are not used for commercial production is due to the very exothermic nature of free radical polymerization and rather limited heat transfer capacity in large-scale reactors. Furthermore, continuous and especially semibatch reaction systems offer the operational flexibility to produce latex products with controlled polymer composition, particle morphology, and particle size distributions. These parameters will have an important influence on the performance properties of latex products. In this chapter, we will focus on the aspects of polymerization mechanisms and kinetics involved in semibatch and continuous emulsion polymerization systems. Those who are interested in the previous studies of semibatch and continuous emulsion polymerization processes should refer to the review articles cited in references 1–6.

7.1 SEMIBATCH EMULSION POLYMERIZATION

7.1.1 Pseudo-Steady-State Polymerization Behavior

Semibatch emulsion polymerization [7–14] is an important process for the manufacture of latex products such as coatings, adhesives and synthetic

Principles and Applications of Emulsion Polymerization, by Chorng-Shyan Chern

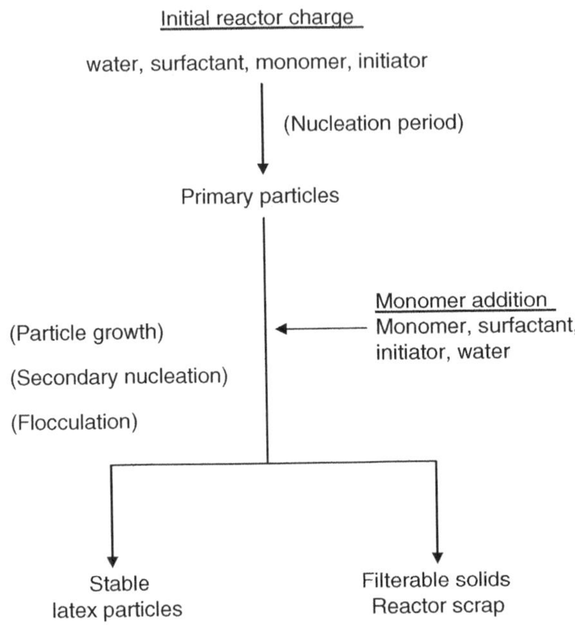

Figure 7.1. Flow chart for a typical semibatch emulsion polymerization process.

elastomers. In addition to its operational flexibility for products with controlled polymer composition, particle morphology, and particle size distribution, the semibatch emulsion polymerization process can remove the enormous heat generated during the reaction. The most striking difference between the semibatch and batch emulsion polymerization processes is that reaction ingredients such as monomer, surfactant, initiator, or water can be added to the semibatch reaction system throughout the polymerization (Figure 7.1). Thus, the residence time distribution of particle nuclei can be broader for semibatch emulsion polymerization. All these features make the semibatch emulsion polymerization mechanisms and kinetics more complicated in comparison with the batch counterpart. Studies dealing with the effects of reaction variables on the polymerization mechanisms and kinetics and colloidal stability are very useful for product and process development.

Wessling [9] studied the reaction kinetics of semibatch emulsion polymerization of relatively hydrophobic monomers such as styrene and derived the following expression for the rate of polymerization at pseudo-steady state (R_p):

$$R_p = F_m/\{1+[N_A/(k_p N_p \mathbf{n})](\mathrm{MW}_m/\rho_m)F_m\} \tag{7.1}$$

where F_m is the monomer feed rate, N_A is Avogadro's constant, k_p is the propagation rate constant, N_p is the number of latex particles per unit volume of water, $\mathbf{n}$ is the average number of free radicals per particle, and MW_m and ρ_m are the molecular weight and density of monomer, respectively. This concise kinetic model was developed based on a mass balance and the assumption that diffusion of monomer molecules into the growing latex particles is instantaneous. Equation (7.1) predicts that at pseudo-steady state, the rate of polymerization approaches the magnitude of the monomer feed rate under monomer-starved conditions [i.e., F_m is very small and the second term in the denominator of Eq. (7.1) can be neglected]. On the other hand, the rate of polymerization at pseudo-steady state is equal to the term $[N_A/(k_pN_p\mathbf{n})](MW_m/\rho_m)$ when the reaction system is operated under monomer-flooded conditions [i.e., F_m is very large and the first term in the denominator of Eq. (7.1) can be neglected]. In other words, under monomer-flooded conditions, the latex particles are saturated with monomer throughout the polymerization and the rate of polymerization at pseudo-steady state is independent of the monomer feed rate until the monomer feed is stopped. This behavior was confirmed experimentally for the semibatch emulsion polymerizations of styrene, methyl acrylate [10], and vinyl acetate [11] and copolymerizations of butyl methacrylate with *n*-butyl acrylate [12], styrene with *n*-butyl acrylate [13], acrylonitrile with *n*-butyl acrylate, and acrylonitrile with styrene [14]. The monomer-flooded condition that shortens the semibatch cycle time should be avoided in large-scale plant production due to the potential hazardous heat transfer problem.

7.1.2 Polymerization Mechanisms and Kinetics

Based on the coagulative nucleation mechanism [15–17], Novak [18] developed a simple model to describe the particle nucleation and growth processes involved in semibatch acrylic emulsion polymerizations. In the model development, it was assumed that the precursor particles (~2 nm in diameter) are first generated by phase separation of the oligomeric radicals in the continuous aqueous phase. These precursor particles, although completely or partially covered with surfactant molecules, are extremely unstable and they tend to aggregate rapidly until a stable primary particle size is achieved. The number of primary particles generated during the particle nucleation period is controlled by the amount of surfactant available to stabilize the oil–water interfacial area created. The remainder of polymerization is simply the growth of these primary particles via the propagation of free radicals with monomer molecules therein. Novak further assumed that there is no particle flocculation and secondary particle nucleation occurring during the monomer addition period. As would be expected, this model predicts that the slope of the $\log(d_{p,f})$ versus $\log(1/W_a)$ plot should be 0.333, where $d_{p,f}$ is the final latex particle diameter and W_a is the weight of anionic surfactant present in the particle nucleation stage.

Chern and Hsu [19] extended the Novak model to the semibatch emulsion copolymerizations of methyl methacrylate and *n*-butyl acrylate stabilized by mixed anionic (sodium dodecyl sulfate) and nonionic (nonylphenol-40 units of ethylene oxide adduct, NP-40) surfactants and initiated by sodium persulfate. They derived the following equation:

$$\log(d_{p,f}) = \frac{1}{3}\log[1/(W_a A_a + W_n A_n)] + \frac{1}{3}\log(N') \tag{7.2}$$

where $N' = 6d_{p,p}^2 W_{p,f}/\rho_p$. The parameter $d_{p,p}$ is the primary particle diameter, $W_{p,f}$ is the final latex particle weight, ρ_p is the polymer density, W_n is the weight of nonionic surfactant used in the particle nucleation period, and A_a and A_n are the particle surface area covered by one gram of anionic and nonionic surfactant, respectively. The concentration of sodium dodecyl sulfate in the initial reactor charge was shown to be the most important parameter in controlling the final latex particle size. The nonionic surfactant in the initial reactor charge is the second most important variable, and it acts as an auxiliary stabilizer. The concentration of sodium dodecyl sulfate in the monomer emulsion feed does not affect the resultant latex particle size very much if the polymerization system remains unsaturated with monomers, and its primary function is to stabilize the growing particles. Other parameters such as the concentrations of the persulfate initiator and monomers in the initial reactor charge and the ratio of methyl methacrylate to *n*-butyl acrylate also have an insignificant influence on the resultant latex particle size. In addition, the resultant particle size increases with increasing the electrolyte concentration in the initial reactor charge or agitation speed (see Chapter 2, Section 2.3). The parameters A_a and A_n are estimated to be 1.41×10^{21} and $4.80 \times 10^{20}\,\text{nm}^2\,\text{g}^{-1}$, respectively, according to the literature [20–22]. Using these values along with Eq. (7.2) gives a slope of 0.410 for the $\log(d_{p,f})$ versus $\log[1/(W_a A_a + W_n A_n)]$ plot, but the experimental data points are somewhat scattered, as shown by the starry data points in Figure 7.2. If the value of A_n is set at $2.38 \times 10^{20}\,\text{nm}^2\,\text{g}^{-1}$, the data points converge rapidly to a straight line with a slope of 0.405, as shown by the circular data points in Figure 7.2. The best-fitted slope (0.405) is slightly greater than that predicted by Eq. (7.2), which indicates that the resultant latex particle sizes are quite sensitive to changes in the levels of mixed surfactants in the initial reactor charge. This implies that the assumption that there is no particle flocculation and secondary particle nucleation occurring during the monomer addition period is invalid for the semibatch emulsion copolymerizations conducted by Chern and Hsu [19]. For example, limited flocculation of latex particles is enhanced when the value of $[1/(W_a A_a + W_n A_n)]$ increases (i.e., the amount of mixed surfactants decreases). As a consequence, the resultant latex particle size is larger than that predicted by the modified Novak model in the region of large $[1/(W_a A_a + W_n A_n)]$ values. This will eventually make the slope of the $\log(d_{p,f})$ versus $\log[1/(W_a A_a + W_n A_n)]$ plot deviate in the increasing direction from the theoretical value (0.333). Secondary particle nucleation (if it occurs) will lead to a reduction in the average latex particle size during

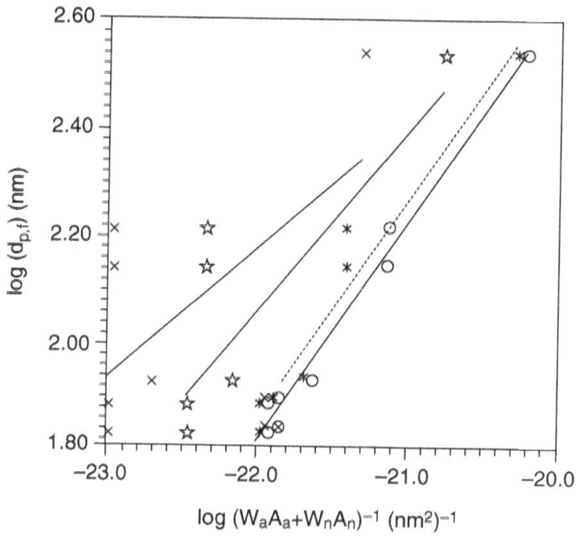

Figure 7.2. Final latex particle diameter versus reciprocal of particle surface area covered by mixed surfactants immediately after particle nucleation. (○) $A_n = 0.238 \times 10^{21}$, slope = 0.405; (×) $A_n = 18.3 \times 10^{21}$, slope = 0.237; (*) $A_n = 4.38 \times 10^{21}$, slope = 0.333; (☆) $A_n = 0.48 \times 10^{21}$, slope = 0.41. The open circular data points represent the best fitting result.

polymerization. A smaller population of latex particles with their surfaces being saturated with surfactant molecules in combination with an abundant supply of surfactant promotes the formation of a second crop of particle nuclei beyond the primary particle nucleation interval. The effect of secondary particle nucleation on the slope of the $\log(d_{p,f})$ versus $\log[1/(W_aA_a + W_nA_n)]$ plot is strongly dependent on the competition for surfactant between the existing latex particles and the particle embryos generated in the continuous aqueous phase or monomer-swollen micelles (if present). Under these circumstances, the modified Novak model cannot be used to distinguish the effect of particle flocculation from that of secondary particle nucleation in semibatch emulsion polymerization.

Gugliotta *et al.* [23] developed a new approach to estimate the monomer conversion and copolymer composition in semibatch emulsion copolymerization systems based on reaction calorimetric measurements. The validity of this technique was confirmed by the semibatch emulsion copolymerizations of both the styrene–*n*-butyl acrylate and vinyl acetate–*n*-butyl acrylate.

Unzueta and Forcada [24] carried out semibatch emulsion copolymerizations of methyl methacrylate and *n*-butyl acrylate stabilized by mixed anionic and nonionic surfactants (sodium dodecyl sulfate and dodecyl polyethoxylate) under the monomer-starved condition. The polymerization system stabilized by nonionic surfactant alone results in a slower rate of polymerization and larger latex particles. At low anionic surfactant concentration, the final number of latex particles per unit volume of water increases with increasing the total surfactant concentration for the polymerization system stabilized by mixed anionic and nonionic surfactants. On the other hand, at high anionic surfactant concentration and a ratio of anionic surfactant to nonionic surfactant between 1:1 and 1:3, a smaller population of latex particles is produced. During the early stage of polymerization, the latex particle size distribution is positively skewed for polymerizations containing no surfactant or low surfactant concentration, narrower and Gaussian-like for polymerizations containing intermediate surfactant concentration, and broader for polymerizations containing high surfactant concentration. Furthermore, latex products with larger particle sizes and narrower particle size distributions were obtained from polymerizations stabilized by mixed anionic and nonionic surfactants compared to the polymerization stabilized by anionic surfactant alone.

Chern et al. [25] focused on the effects of various reaction variables on the colloidal stability of acrylic latex particles in the course of polymerization. Sodium dodecyl sulfate was used as the sole surfactant. The amount of coagulum produced by the intensive coagulation of latex particles is greatly reduced when the level of sodium dodecyl sulfate in the monomer emulsion feed increases. The increased level of sodium dodecyl sulfate in the initial reactor charge leads to an increase in the volume change of latex particles due to the limited particle flocculation later in the polymerization process. The larger the volume change of latex particles, the stronger the limited particle flocculation. The larger the ratio of methyl methacrylate to *n*-butyl acrylate in the copolymer, the greater the amount of coagulum produced. This is closely related to the polarity of latex particle surfaces formed during polymerization. The higher the polarity of latex particle surfaces, the lower the amount of surfactant that can be adsorbed on the oil–water interface. As a result, the latex particles comprising more monomeric units of methyl methacrylate exhibit inferior colloidal stability. As expected, both the amount of coagulum and the volume change of latex particles increase with increasing the electrolyte concentration. The agitation speed (500–800 rpm) shows an insignificant influence on the particle coagulation process. Furthermore, latex particles lose their colloidal stability rapidly above 40% total solids content due to the crowding effect. Chern and Lin [26] illustrated that latex particles are capable of maintaining appreciable colloidal stability via the limited particle flocculation process during the semibatch surfactant-free emulsion polymerization of *n*-butyl acrylate initiated by sodium persulfate. Nevertheless, in contrast to the polymerization system stabilized by sodium dodecyl sulfate [25], intensive agitation (e.g., 800 rpm in a 1-dm^3 reactor) results in a significant amount of coagulum, in addition to the limited particle flocculation effect experienced in the semibatch surfactant-free emulsion polymerization system. The average size of latex particles first increases to a maximum and then decreases when the initiator concentration increases. The polymerization with the lowest level of initiator exhibits the worst colloidal stability because of the insufficient supply of sulfate end-groups to stabilize the growing latex particles. However, a significant amount of coagulum can be observed for the polymerization with the highest level of initiator due to the ionic strength effect (see Chapter 2, Section 2.3). The optimal colloidal stability occurs at a point close to an initiator level of 0.19%. The effects of sodium bicarbonate (buffer) on the latex particle size and colloidal stability are significant, again owing to the ionic strength effect. The latex products based on this type of recipes may find potential applications in pressure sensitive adhesives.

To overcome the poor water resistance generally experienced with waterborne polymers, the level of surfactant used in emulsion polymerization must be minimized. However, the colloidal stability of latex particles can be greatly reduced and a significant amount of coagulum can form during the monomer addition period, which is not acceptable in the plant production. Furthermore, latex particles can grow in size by relatively mild particle agglomeration in order to maintain appreciable colloidal stability, in addition to polymerizing the imbibed monomer in the particles. The improved latex stability is attributed to the decreased particle surface area and, hence, the increased particle surface charge density associated with such a limited particle flocculation process. The limited particle flocculation process makes the task of particle size control even more difficult. This is a critical issue because control of the particle size and particle size distribution is the key to guaranteeing the quality of latex products. To get around the dilemma between satisfactory product performance properties and smooth plant production, a small amount of functional monomers such as acrylic acid can be incorporated into the emulsion polymers to greatly enhance their colloidal stability. The ionized carboxyl group that is chemically incorporated into the emulsion polymer can increase the particle surface charge density and, therefore, increase the electrostatic repulsion force among the interactive particles. Chern and Lin [27] studied the effects of functional monomers (acrylic acid, methacrylic acid, and 2-hydroxyethyl methacrylate) on the semibatch emulsion polymerization of *n*-butyl acrylate stabilized by the mixed surfactants of sodium dodecyl sulfate and nonylphenol-40 units of ethylene oxide aduct and initiated by sodium persulfate. The experimental results show that the final latex particle size decreases with increasing the concentration of functional monomers. Among the functional monomers investigated, acrylic acid was the most efficient in nucleating and then stabilizing the latex particles. The resultant latex particle size first decreases to a minimum and then increases with increasing the concentration of sodium bicarbonate (a neutralizing agent). The optimal sodium bicarbonate concentration for achieving the smallest latex particle size occurs at a point close to 0.15–0.29%. Furthermore, the aqueous phase polymerization plays an important role during the particle nucleation period. The colloidal stability of latex particles during the semibatch emulsion polymerization of *n*-butyl acrylate in the presence of 0–10% acrylic acid stabilized by sodium dodecyl sulfate and initiated by sodium persulfate was also investigated [28]. Incorporation of 5% acrylic acid into emulsion polymers greatly improves their colloidal stability. Both the particle coagulation and secondary particle nucleation processes cause a significant deviation of the particle nucleation and growth involved in the semibatch emulsion polymerizations of *n*-butyl acrylate in the presence of acrylic acid from the Novak model [18]. In an attempt to improve the colloidal stability and, therefore, gain a better control of the latex particle size, the effects of incorporating a small quantity of acrylic acid or methacrylic acid into the semibatch surfactant-free emulsion polymerizations of *n*-butyl acrylate were investigated [29]. The final latex particle size first increases to a maximum occurring at around 2% acrylic acid and then decreases with increasing the concentration of acrylic acid. This can be attributed to the effect of polyelectrolyte (oligomers comprising acrylic acid and *n*-butyl acrylate) generated in the particle nucleation period. In addition, the resultant latex particle size is independent of the type of carboxylic monomers. As expected, a significant fraction of the monomeric units of acrylic acid is present near the polymer particle surface layer. On the other hand, the less hydrophilic monomeric units of methacrylic acid are distributed more uniformly in the particles.

Alternatively, conventional surfactants used to nucleate and stabilize latex particles can be replaced by reactive surfactants (or polymerizable surfactants). In addition to the basic surfactant properties (e.g., lowering of surface tension and formation of micelles in water), the surfactant containing one carbon–carbon double bond per molecule can be chemically incorporated onto the particle surfaces during emulsion polymerization. Thus, the water sensitivity of latex products arising from the immobilized surfactant molecules near the particle surface layer can be minimized. However, latex products with surfactant molecules being immobilized on the polymer particle surfaces show very high surface tension and, consequently, result in poor wetting characteristics. Thus, a large amount of wetting agent (HLB = 7–9, see Chapter 2, Section 2.2.2), which effectively lowers the surface tension, is normally required to adequately spread the emulsion polymer on a substrate with relatively low surface energy. This may cause some detriment to the optical properties of the coating films.

Chern and Chen [30] studied the effect of the reactive surfactant, sodium dodecyl allyl sulfosuccinate, on the semibatch emulsion polymerization of *n*-butyl acrylate initiated by sodium persulfate. Sodium dodecyl allyl sulfosuccinate plays a similar role in the particle nucleation and growth stages to the conventional sodium dodecyl sulfate. The final number of latex particles per unit volume of water is proportional to the concentration of sodium dodecyl allyl sulfosuccinate in the initial reactor charge (the most important parameter with regard to particle nucleation) to the 0.72–0.80 power. The saturated particle surface area occupied by one molecule of sodium dodecyl allyl sulfosuccinate is $0.36\,nm^2$ for the poly(*n*-butyl acrylate) particles prepared by the surfactant-free emulsion polymerization. Furthermore, the value of the saturated particle surface area increases with increasing the particle surface polarity for the acrylic latexes stabilized by the reactive surfactant. The kinetic studies indicated that an induction period or even a complete inhibition of the polymerization is observed for the experiments with relatively high sodium dodecyl allyl sulfosuccinate concentration or low sodium persulfate concentration [31]. This is attributed to the intensive chain transfer of free radicals to the reactive surfactant molecules.

Xu and Chen [32] prepared two polymerizable surfactants, sodium 4-(ω-acryloyloxyalkyl)oxy benzene sulfonate with the alkyl chain length equal to 8 or 10, and used them to stabilize the semibatch emulsion copolymerization of butyl methacrylate. A redox initiator system of ammonium persulfate and tetramethylethylenediamine was used to start the polymerization at room temperature. The latex particle size increases continuously, whereas the number of particles per unit volume of water remains relatively constant with the progress of polymerization. This is attributed to the predominant micellar nucleation mechanism. X-ray photoelectron spectroscopy data show that polymerizable surfactant molecules are preferably located near the latex particle surface layer.

The seeded technique (i.e., the number of latex particles per unit volume of water in the reaction system is known immediately before the start of monomer feed) has been widely used to study the particle growth mechanisms in the semibatch emulsion polymerization in order to avoid the complicated effects caused by particle nucleation. Furthermore, from the industrial point of view, semibatch emulsion polymerization in the presence of a constant seed particle concentration results in better quality control of latex products. Chern et al. [33] studied particle growth mechanisms involved in the semibatch surfactant-free emulsion polymerization of *n*-butyl acrylate in the presence of poly(*n*-butyl acrylate) seed particles. The experimental results show that limited particle flocculation, often observed in the polymerization containing no seed particles, did not take place in this work. This is attributed to the extensive formation of coagulum during polymerization. Agitation speed was the most important parameter in determining the level of total scrap produced, followed by the initiator (sodium persulfate) concentration, monomer feed rate, and then buffer concentration. Furthermore, levels of coagulum for the polymerizations using carboxylic seed particles were much lower than those for the polymerizations using ordinary poly(*n*-butyl acrylate) seed particles. Again, limited particle flocculation did not occur during these semibatch emulsion polymerizations. In this case, nucleation of a second crop of primary particles during the monomer addition period was confirmed. Chern et al. [34] prepared and characterized a series of highly carboxylated seed latexes comprising methacrylic acid or acrylic acid, dodecyl mercaptan, and methyl methacrylate. Dodecyl mercaptan was used to regulate the polymer molecular weight. These neutralized latexes were then used as the seed particles for the subsequent semibatch surfactant-free emulsion polymerization of *n*-butyl acrylate. The seed latexes containing 50 wt% methacrylic acid or acrylic acid at pH 8–9 were very effective for this purpose due to the increased particle surface charge density.

Sajjadi and Brooks [35] showed that the concentration of monomer in the initial reactor charge has a significant influence on the final number of latex particles per unit volume of water in the semibatch emulsion polymerization of *n*-butyl acrylate. The polymerization system operated under the monomer-starved condition promotes nucleation of latex particles due to retarded particle growth. As a result, more particle nuclei are generated for the polymerization when the concentration of monomer in the initial reactor charge is below its critical value. The largest population of latex particles was obtained from the polymerization without any monomer in the initial reactor charge. In addition, the pseudo-steady-state rate of polymerization is only slightly dependent on the number of latex particles per unit volume of water. They also demonstrated that bimodal particle size distributions can be obtained from the semibatch emulsion polymerization of *n*-butyl acrylate using the monomer emulsion feed with a constant rate and no surfactant in the initial reactor charge [36]. Variations in the monomer emulsion feed rate, concentrations of monomer, surfactant, and initiator, and their distribution between the initial reactor charge and the feed stream were shown to have significant effects on the resultant latex particle size distribution. Furthermore, the rate of secondary particle nucleation is inversely proportional to that of primary particle nucleation. The particle size distribution of latex products became broader when the monomer feed rate was decreased. The aqueous phase polymerization plays an important role in the secondary particle nucleation process. The authors then studied the semibatch emulsion polymerization of *n*-butyl acrylate with neat monomer feed [37]. Sodium dodecyl sulfate and potassium persulfate were used as the surfactant and initiator, respectively. The parameters investigated included monomer feed rate, concentrations of surfactant and initiator, and temperature. At pseudo-steady state, the polymerization system followed the Wessling model [9]. The time required to reach the near pseudo-steady state increased with increasing the monomer feed rate for polymerizations when the polymer particles were not swollen with monomer, whereas it decreases with increasing monomer feed rate for the polymerization when the particles were swollen with monomer. Nevertheless, the time required to reach the real steady state was always minimal at the lowest monomer feed rate regardless of the seed particle compositions.

Latex products with high solids content and acceptable rheological properties are desirable in many industrial applications such as adhesives, coatings, and caulks and sealants. Such concentrated polymer colloids are generally manufactured by semibatch emulsion polymerization processes and characterized by the extremely efficient packing of the polymer particles with broad particle size distributions. Thus, how to effectively manipulate the complex reaction variables to precisely control the latex particle size distribution is crucial for the related product development program.

Chern et al. [38] demonstrated that injecting sodium dodecyl sulfate into the reaction medium induces a second crop of tiny primary particles in the semibatch seeded emulsion polymerization of acrylic monomers. As a result, latex products with bimodal particle size distributions are obtained. The concentration of seed particles is the most important parameter that controls the latex particle size distribution, followed by the time when the surfactant is injected into the reactor. Retarded secondary particle nucleation during the monomer addition period is achieved when the concentration of seed particles or the time when the surfactant is injected into the reactor increases. The total seed particle surface area is shown a very useful parameter for manipulating the resultant latex particle size and particle size distribution. Schneider *et al.* [39] proposed two methods for preparing high solids content (over 65%) latex products with bi- and trimodal particle size distributions in the semibatch emulsion copolymerization of methyl methacrylate, *n*-butyl acrylate, and a small amount of acrylic acid. The key parameter in determining the colloidal stability of the latex particles, formation of coagulum, and rheology is the desirable particle size distribution that can be modified effectively by secondary particle nucleation. Nucleation of a third crop of particles for the latex product with a trimodal particle size distribution is best accomplished with mixed anionic and nonionic surfactants because anionic surfactant alone generally results in detrimental changes in the particle size distribution as a consequence of excessive particle flocculation. Boutti et al. [40] prepared high-solids (over 70%) latex products without resorting to the use of intermediate seed particles in the semibatch emulsion copolymerization of acrylic monomers. An electrostatically neutral initiator (hydrogen peroxide) and relatively low levels of mixed anionic and nonionic surfactants were used to start the free radical polymerization in the initial stage in order to avoid stabilizing small particle nuclei. This was followed by the use of a persulfate initiator in the second half of the semibatch emulsion polymerization process. In this manner, a desirable population of small particle nuclei is generated and adequately stabilized.

Sebenik and Krajnc [41] studied the effect of the soft segment (polyester polyol) chain length on the reaction kinetics of the semibatch emulsion copolymerization of methyl methacrylate, *n*-butyl acrylate, and a small amount of acrylic acid in the presence of polyurethane seed particles. The water-borne polyurethane seed particles were prepared by condensation reactions of isophorone diisocyanate, polyneopentyladipate (M_n = 1000 or 2000 g mol^{-1}), and dimethylol propionic acid. The rigidity of polyurethane is controlled by varying the soft segment chain length. The weight ratio of acrylic monomers to polyurethane was kept constant at 1 : 1. The polyurethane seed particles containing the higher-molecular-weight polyol were swollen with acrylic monomers to a greater extent. In the pseudo-steady state, the number of latex particles per unit volume of water remains relatively constant, and it is comparable to the number of polyurethane seed particles present in the initial reactor charge.

The polymerization system shows intermediate behavior between Smith–Ewart Cases 1 and 2.

7.1.3 Mathematical Modeling Studies

Chern [42] developed a mechanistic model based on diffusion-controlled reaction mechanisms to predict the kinetics of the semibatch emulsion polymerization of styrene. Reasonable agreement between the model predictions and experimental data available in the literature was achieved. Computer simulation results showed that the polymerization system approaches Smith–Ewart Case 2 kinetics (**n** = 0.5) when the concentration of monomer in the latex particles is close to the saturation value. By contrast, the polymerization system under the monomer-starved condition is characterized by the diffusion-controlled reaction mechanisms (**n** > 0.5). The author also developed a model to predict the effect of desorption of free radicals out of the latex particles on the kinetics of the semibatch emulsion polymerization of methyl acrylate [43]. The validity of the kinetic model was confirmed by the experimental data for a wide range of monomer feed rates. The desorption rate constant for methyl acrylate at 50 °C was determined to be $4 \times 10^{-12}\,cm^2\,s^{-1}$.

Wang et al. [44] used the Langmuir site adsorption model in combination with surface tension measurements to control the level of surfactant (sodium dodecylbenzene sulfonate) fed to the semibatch seeded emulsion copolymerization of methyl methacrylate and *n*-butyl acrylate with the goal to avoid secondary particle nucleation or coagulation. The Langmuir site adsorption model was developed to relate the surface tension to surfactant concentration in the polymerization system from first principles. It was used to predict the partitioning of the added surfactant and mobile *in situ* surfactant between the continuous aqueous phase and the particle–water interface in the presence of anchored sulfate end-groups originating from the persulfate initiator. For example, monodisperse latex products with average particle diameters of 0.5–3 μm were obtained when the surface tension was maintained at 45–57 dyn cm^{-1}.

Chern and Kuo [45] developed a kinetic model to simulate the shear-induced particle coagulation process during the semibatch seeded emulsion polymerization of acrylic monomers. DLVO theory was used to calculate the total potential energy barrier against the coagulation of latex particles. The coagulation rate constant was postulated to be proportional to an exponential function of the relatively mean free path length between two latex particles. Agreement between the model predictions and experimental data was good. The model was then used to study the effects of important reaction variables such as the total solids content, seed particle size and concentration, initiator concentration, and surfactant feed profile on the colloidal stability of the latex particles nucleated in a semibatch reactor.

Based on the coagulative particle nucleation mechanism, a two-step model was developed for the semibatch surfactant-free emulsion polymerization of *n*-butyl acrylate in the absence or presence of a small amount of acrylic acid [46]. During Stage 1, precursor particles are generated by phase separation of oligomeric radicals in the continuous aqueous phase, followed by the limited flocculation of unstable precursor particles to form stable primary particles. During the early part of Stage 2, the rate of limited flocculation of precursor particles to form primary particles is slower than that of coagulation of primary particles. This results in a decreased particle concentration with time. Later, the particle concentration starts to level off and finally reaches a steady value. The model predicts the experimental data of particle concentration and particle size with the progress of polymerization reasonably well.

Sajjadi [47] developed two mechanistic models for the particle nucleation process involved in the semibatch emulsion polymerization of styrene under the monomer-starved condition. In the first model, Smith–Ewart theory was extended to take into account the particle nucleation under the monomer-starved condition. The number of latex particles per unit volume of water is proportional to the surfactant concentration, the rate of initiator decomposition, and the rate of monomer addition, respectively, to the 1.0, 2/3, and –2/3 powers. The second model considers the aqueous phase polymerization kinetics and its effect on the efficiency of free radical capture by the monomer-swollen micelles. This model is capable of predicting some features of the particle nucleation process.

Immanuel et al. [48] developed a population balance model for the nonionic surfactant-stabilized emulsion copolymerization of vinyl acetate and *n*-butyl acrylate. The initiator package included *t*-butyl hydrogen peroxide and sodium formaldehyde sulfoxylate. The model takes into account the effects of the nucleation, growth, and flocculation of latex particles on the evolution of the latex particle size distribution. Recently, Sajjadi and Yianneskis [49] studied the particle nucleation and growth mechanisms involved in the semibatch emulsion polymerization reaction system operated under the monomer-starved condition. Among the parameters investigated, the monomer feed rate together with the level of monomer in the initial reactor charge were the primary factors that control the particle nucleation process. Reducing the rate of particle growth prolongs the particle nucleation period and, therefore, slows down the rate of micelle depletion. As a result, a larger population of latex particles is established during polymerization.

7.2 CONTINUOUS EMULSION POLYMERIZATION

7.2.1 General Features of Continuous Emulsion Polymerization Processes

Continuous emulsion polymerization processes have been widely used in the large-scale production of latex products. For example, styrene–butadiene rubber latex products are manufactured by such processes consisting of a number (5–15) of stirred tank reactors in series. Different reactor configurations were also developed for continuous emulsion polymerizations [50]. Several advantages and disadvantages are summarized as follows [51]:

Advantages

(a) Production of large-volume latex products at lower costs.
(b) Consistent quality of latex products.
(c) Full utilization of heat transfer capacity.
(d) Fewer problems with coagulation of latex particles and formation of wall polymer.

Disadvantages

(a) Less flexibility in terms of the operation and control of product characteristics.
(b) Potential production of off-specification emulsion polymer during the start-up step or change over of latex products.
(c) Difficulties with the direct development of continuous emulsion polymerization processes based on the information obtained from batch and semibatch experiments.

Reactor configurations involved in continuous emulsion polymerization include stirred tank reactors, tubular reactors, pulsed packed reactors, Couett–Taylor vortex flow reactors, and a variety of combinations of these reactors. Some important operational techniques developed for continuous emulsion polymerization are the prereactor concept, start-up strategy, split feed method, and so on. The fundamental principles behind the continuous emulsion polymerizations carried out in the basic stirred tank reactor and tubular reactor, which serve as the building blocks for the reaction systems of commercial importance, are the major focus of this chapter.

Just like a batch reactor, all the reacting species in an idealized tubular reactor have exactly the same residence time. Thus, the mechanisms and kinetics presented in Chapters 3 and 4 are also applicable to emulsion polymerization carried out in a tubular reactor. On the other hand, the feed stream introduced into a continuous stirred tank reactor at any given time becomes completely mixed with the reaction mixture already present in the reaction system. As a result, a distribution of residence times of the material within a continuous stirred tank reactor is achieved. In other words, some of the recipe ingredients entering the continuous stirred tank reactor may leave it almost immediately because material is continuously withdrawn from the reactor. In contrast, other recipe ingredients may remain in the reactor almost forever because all the material is never removed from the reactor at one time. Many of the reaction species leave the reactor after spending a period of time somewhere in the vicinity of the mean residence time. The distribution of residence

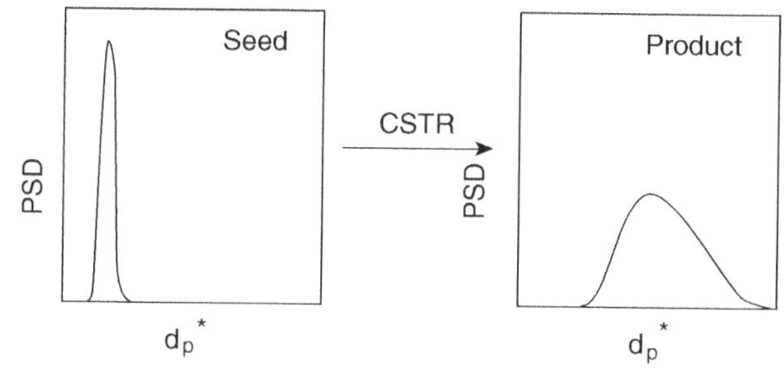

Figure 7.3. A schematic representation for a continuous emulsion polymerization process, in which the relatively monodisperse particle size distribution of seed latex particles introduced into a continuous stirred tank reactor becomes broader at the exit of the reactor.

times in a continuous stirred tank reactor is expected to significantly affect its performance. As an example, the relatively monodisperse particle size distribution of seed latex particles introduced into a continuous stirred tank reactor becomes broader at the exit of the reactor, as schematically shown in Figure 7.3.

One unique but normally undesirable feature of continuous emulsion polymerization carried out in a stirred tank reactor is reactor dynamics. For example, sustained oscillations (limit cycles) in the number of latex particles per unit volume of water, monomer conversion, and concentration of free surfactant have been observed in continuous emulsion polymerization systems operated at isothermal conditions [52–55], as illustrated in Figure 7.4a. Particle nucleation phenomena and gel effect are primarily responsible for the observed reactor instabilities. Several mathematical models that quantitatively predict the reaction kinetics (including the reactor dynamics) involved in continuous emulsion polymerization can be found in references 56–58. Tauer and Muller [59] developed a kinetic model for the emulsion polymerization of vinyl chloride in a continuous stirred tank reactor. The results show that the sustained oscillations depend on the rates of particle growth and coalescence. Furthermore, multiple steady states have been experienced in continuous emulsion polymerization carried out in a stirred tank reactor, and this phenomenon is attributed to the gel effect [60, 61]. All these factors inevitably result in severe problems of process control and product quality.

To resolve this instability problem, adopting a feed stream of seed latex particles [62] or installing a continuous tubular reactor, which generates seed particles, upstream of the continuous stirred tank reactor [53] have been proved quite effective (Figure 7.4b). For the latter approach, small latex particles form as a seed latex before the reacting stream enters the continuous stirred tank reactor when the monomer conversion at the exit of the tubular reactor is maintained at an adequate level. As a result, the continuous emulsion polymerization system can be operated at a stable steady state. The work of Nomura and Harada [54] also suggests that a tube-stirred tank reactor series

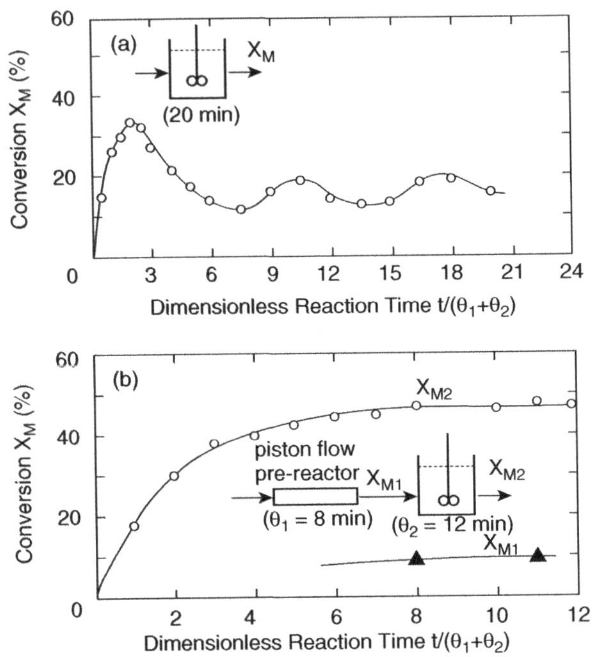

Figure 7.4. Monomer conversion as a function of the dimensionless reaction time for the emulsion polymerization of vinyl acetate in **(a)** a single continuous stirred tank reactor and **(b)** a tube-stirred tank reactor series [54].

may be an optimal reactor design for the continuous production of latex products.

Before leaving this section, a brief discussion on the tubular reactor system with a recirculation loop configuration, which exhibits some interesting features, is worthwhile. First, this reaction system generally has an extremely large surface area available for heat transfer, which permits very fast polymer reactions to take place therein. As a result, the total reactor volume of this type can be much smaller than other competitive polymerization systems. This implies that fewer raw materials are present in the reactor; thus, inevitable waste produced during start-up and product changes can be considerably less than that of a standard continuous stirred tank reactor. If the recirculation rate of the reaction medium in the loop is significantly greater than the flow-through rate, which is usually the case, the tubular reactor with a recirculation loop is very similar to a continuous stirred tank reactor in terms of residence time distribution and other related polymerization parameters.

Those who are interested in continuous emulsion polymerization should refer to representative review articles [1, 6, 50, 51].

7.2.2 Particle Nucleation and Growth Mechanisms

The pioneering work of Gershberg and Longfield [52] deals with the steady-state number of polystyrene particles per unit volume of water nucleated in a stirred tank reactor or a series of stirred tank reactors based on the conventional Smith–Ewart theory (see Chapter 3, Section 3.1.1), as shown by the following balance equation.

$$dN_p/dt = \rho_i[A_m/(A_m + A_p)] - N_p/\theta = 0 \tag{7.3}$$

where ρ_i is the rate of generation of free radicals in the continuous aqueous phase, A_m and A_p are the total micelle surface area and total particle surface area per unit volume of water, respectively, and θ is the mean residence time of a single stirred tank reactor. Substitution of the total latex particle surface area (a_sS_0, where a_s is the particle surface area occupied by unit weight of the adsorbed surfactant and S_0 is the weight of surfactant initially present in the feed) for $A_m + A_p$ in Eq. (7.3) yields

$$N_p = \rho_i[1-(A_p/a_sS_0)] \tag{7.4}$$

The residence time distribution function $E(t)$ of the growing latex particles in a perfectly mixed stirred tank reactor is defined as follows:

$$\begin{aligned} E(t)dt &= dN_p/N_p \\ &= (1/\theta)\exp(-t/\theta)dt \end{aligned} \tag{7.5}$$

The parameter A_p can be obtained from the following integral:

$$\begin{aligned} A_p &= \int_0^{N_p} a_p dN_p \\ &= (36\pi)^{1/3}\int_0^{N_p} v_p^{2/3} dN_p \\ &= (36\pi)^{1/3} N_p \int_0^{\infty} (\mu t)^{2/3} E(t)\,dt \end{aligned} \tag{7.6}$$

where a_p is the surface area of a latex particle, v_p is the volume of a particle, and $\mu = dv_p/dt$ is the volumetric growth rate for particle nuclei. Finally, substituting Eq. (7.6) into Eq. (7.4) yields an analytical solution describing the particle nucleation process involved in a perfectly mixed stirred tank reactor.

$$N_p = \rho_i\theta/[1+(4.36\rho_i\mu^{2/3}\theta^{5/3}a_sS_0)] \tag{7.7}$$

For constant surfactant and initiator concentrations and particle growth rate, the number of latex particles per unit volume of water is linearly proportional to mean residence time when θ approaches zero and the second term in the denominator can be neglected. In other words, at small θ values, N_p increases linearly with increasing θ. On the other hand, the relationship $N_p \sim \theta^{-2/3}$ holds when θ becomes very large. At large θ values, N_p decreases with increasing θ, as illustrated in reference 52. This implies that a maximal number of latex particles per unit volume of water ($N_{p,\max}$) exists at the optimal mean residence time ($\theta_{\max}$), as shown by the following equations [63, 64]:

$$\theta_{\max} = 0.53(a_s S_0/\rho_i \mu^{2/3})^{3/5} \tag{7.8}$$

$$N_{p,\max} = 0.21(\rho_i/\mu)^{2/5}(a_s S_0)^{3/5} \tag{7.9}$$

It is also interesting to note that the maximal number of latex particles per unit volume of water ($N_{p,\max}$) that can be achieved in a continuous emulsion polymerization system is 58% of that ($N_{p,\mathrm{I}}$) nucleated in the batch counterpart with exactly the same recipe and temperature [cf. Eq. (7.9) with Eq. (3.5)]. In addition, the optimal mean residence time ($\theta_{\max}$) is 83% of the time (t_I) at which particle nucleation stops as a result of complete depletion of monomer-swollen micelles in a batch emulsion polymerization system (see Chapter 1, Section 1.1.1).

DeGraff and Poehlein [65] proposed a steady state kinetic model for predicting the particle nucleation rate, particle growth rate and particle size distribution of the continuous emulsion polymerization of styrene in a stirred tank reactor. An important contribution of this work is the introduction of the concept of residence time distribution to continuous emulsion polymerization theory. Poehlein et al. [66–71] modified the monodisperse particle size distribution analysis of Ugelstad et al. [72] and then applied to a single continuous stirred tank reactor with a feed stream of seed latex particles in order to eliminate particle nucleation in the stirred tank reactor. The following four dimensionless groups are used to predict the rate of polymerization, particle size distribution, and polymer molecular weight distribution of the latex product in the continuous stirred tank reactor effluent.

$$\alpha_c' = \rho_i <v_s>/(k_{t,p} N_p) \tag{7.10}$$

$$\gamma = k_{\mathrm{des}}' <v_s>^{1/3}/k_{t,p} \tag{7.11}$$

$$Y_c = 2 N_p k_{t,p} k_{t,w} \Big/ \left\{ \left[4\pi D_w N_p (3<v_s>/(4\pi))^{1/3} \right]^2 <v_s> \right\} \tag{7.12}$$

$$\beta = <v_s>/(\theta K_1 [\mathrm{M}]_p) \tag{7.13}$$

where α_c', γ, Y_c, and β are related to the absorption of free radicals by the latex particles, desorption of free radicals out of the particles, bimolecular termination of free radicals in the continuous aqueous phase, and mean residence time in the continuous stirred tank reactor, respectively. The kinetic parameter $\langle v_s \rangle$ is the average volume of the seed latex particles, $k_{t,p}$ and $k_{t,w}$ are the bimolecular termination rate constants in the particle phase and in the water phase, respectively, k_{des}' is the desorption rate constant that is independent of the particle size, D_w is the diffusion coefficient of monomeric radicals in water, and $[\mathrm{M}]_p$ is the concentration of monomer in the particles. K_1 is a constant that is related to the latex particle growth rate according to the following expression:

$$dv_p^*/d\tau = \theta K_1 [\mathrm{M}]_p \mathbf{n}/<v_s> \tag{7.14}$$

where v_p^* is the dimensionless latex particle volume defined as $v_p/\langle v_s \rangle$ (v_p = particle volume), and τ is the dimensionless time defined as t/θ (t = real time).

The rate of polymerization in a continuous stirred tank reactor can be written as

$$R_p = k_p [\mathrm{M}]_p (<\mathbf{n}> N_p / N_A) \tag{7.15}$$

$$<\mathbf{n}> = \int \mathbf{n} U'(v_p^*) dv_p \tag{7.16}$$

$$\mathbf{n} = (8\alpha_c/<v_p^{*1/3}>)^{1/2} (v_p^{*2/3}/4)$$
$$\{I_\gamma v_{p1/3}^*[(8\alpha_c/<v_p^{*1/3}>^{1/2} v_p^{*2/3})]/I_\gamma v_{p1/3}^* - 1[(8\alpha_c/v_p^{*1/3})^{1/2} v_p^{*2/3}]\} \tag{7.17}$$

$$\alpha_c = \alpha_c' + \gamma <\mathbf{n}/v_p^{*2/3}> - Y_c \alpha_c^2/<v_p^{*1/3}>^2 \tag{7.18}$$

where $\langle \mathbf{n} \rangle$ is the average number of free radicals per particle for the latex particles with a broad particle size distribution arising from the different ages of the particles in the reactor, $U'(v_p^*)(= \beta e^{-\tau}/\mathbf{n})$ is the particle volume density function, and N_A is the Avogadro number. Equations (7.17) and (7.18) are analogous to the Stockmayer–O'Toole theory [73, 74] for monodisperse latex particles produced in a batch reactor (see Chapter 4, Section 4.1.2).

Molecular weight and molecular weight distribution are important parameters that govern physical and end-use properties of polymer products such as mechanical strength and processibility. The pioneering work on modeling the molecular weight distribution of emulsion polymer was carried out by Katz and co-workers [75–78]. They took into account the key feature of the compartmentalization of free radicals among the discrete submicron latex particles and adopted both stochastic and deterministic approaches to model the molecular weight distribution of polymer generated in batch emulsion polymerization. DeGraff and Poehlein [65] applied the stochastic model of Katz et al. [77] to the emulsion polymerization carried out in a continuous stirred tank reactor. In this mathematical model, the rate of absorption of free radicals by the latex particles is a function of particle size. The model predicts that the polydispersity index of the molecular weight distribution of polymer developed in continuous emulsion polymerization in a stirred tank reactor has a value of 4.84, which is much larger than the value (2.0) that would be expected in the emulsion polymerization system comprising monodisperse latex particles when the absorption of free radicals by the particles follows the constant flux model. In contrast, a value between 2.0 and 4.84 for the polydispersity index of the molecular weight distribution of polymer can be achieved when the continuum diffusion model is employed.

Lichti et al. [79, 80] developed a comprehensive model to predict the molecular weight and molecular weight distribution of linear polymer chains formed in batch emulsion polymerization. This mechanistic model takes into consideration most important microscopic kinetic events such as the absorption, desorption, and bimolecular termination of free radicals, chain transfer reactions, and reabsorption and cross-termination of the desorbed free radicals. Lee [70] then extended the model of Lichti et al. [79, 80] to predict the molecular weight and molecular weight distribution of polymer chains produced in the continuous emulsion polymerization of styrene. The absorption and desorption of free radicals that are dependent on the particle size were incorporated into the model. Computer simulations were carried out to study the effects of chain transfer of free radicals to chain transfer agents or monomer and desorption of free radicals out of the latex particles on the instantaneous molecular weight and molecular weight distribution of latex product obtained from a seed-fed continuous stirred tank reactor operated at steady state. It was shown that the calculated number–average molecular weight of polymer is inversely proportional to the degree of chain transfer reactions. Furthermore, the polydispersity index of molecular weight distribution of polymer is very close to 2.0, which is as would be expected for a polymerization system controlled by chain transfer reactions and/or desorption of free radicals. These results are consistent with the work of Lichti et al. [79, 80]. Interesting enough, particle size and desorption of free radicals that have significant effects on the emulsion polymerization kinetics show an insignificant influence on the molecular weight and molecular weight distribution of polymer as compared to that of chain transfer reactions. This confirms the fact that using chain transfer agents is the most effective method to regulate the molecular weight of emulsion polymer. Nevertheless, as discussed above, the molecular weight distribution of polymer is relatively insensitive to changes in the concentration of chain transfer agent and it approaches a lower limit (2.0) as the concentration of chain transfer agent is increased to a high enough level.

7.3 DEVELOPMENT OF COMMERCIAL CONTINUOUS EMULSION POLYMERIZATION PROCESSES

It is not straightforward to successfully manufacture a particular latex product, which is generally developed in a laboratory batch or semibatch reactor, in a commercial continuous emulsion polymerization system (e.g., a continuous stirred tank reactor). This is simply because the characteristics of continuous stirred tank reactors are dramatically different from those of batch and semibatch reactors. As a consequence, the particle nucleation process and kinetics experienced in batch or semibatch emulsion polymerization systems cannot be directly applied to continuous systems consisting of stirred tank reactors.

Poehlein [81] identified major problems encountered with the development of continuous emulsion polymerization processes. It was shown that the development of commercial continuous emulsion polymerization processes involves the consideration of many factors associated with process design and product quality. These factors include the effects of inhibitor, polymerization rate, particle size distribution, copolymer composition, addition strategy of feed streams, unsteady-state operation, and reactor design on continuous emulsion polymerization processes. The author then used a two-continuous stirred tank reactor series to elucidate key continuous emulsion polymerization mechanisms and generate the knowledge necessary for the development of commercial continuous processes.

Taking latex particle size distribution as an example, nucleation of latex particles in the early stage of an ideal batch emulsion polymerization is normally very fast. Therefore, the total particle surface area generated is large enough to capture the free surfactant molecules and particle nucleation ceases quite early in the polymerization (~2–10% monomer conversion). Latex particles thus formed would have approximately identical ages at the end of an ideal batch polymerization (Figure 7.5a). The particle size distribution of these

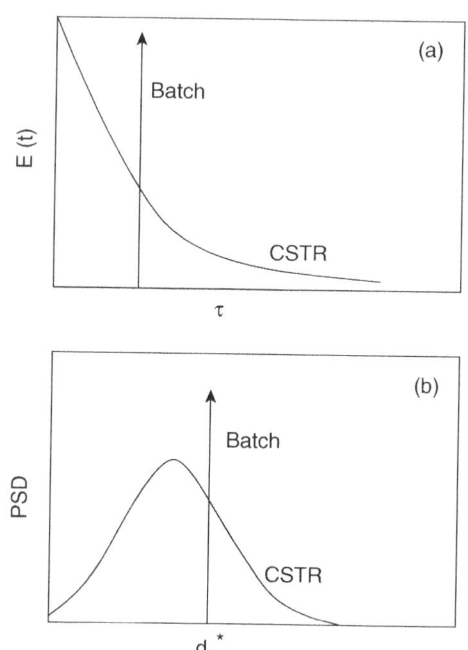

Figure 7.5 Residence time distribution function as a function of the dimensionless reaction time for an ideal batch or continuous stirred tank reactor. CSTR, PSD, and d_p^* represent continuous stirred tank reactor, particle size distribution function, and dimensionless particle diameter, respectively.

latex particles is extremely narrow provided that the effects of particle flocculation, secondary particle nucleation and stochastic differences are insignificant (Figure 7.5b). On the other hand, the residence time distribution function [$E(t)$] of the growing latex particles in a perfectly mixed stirred tank reactor, which is defined as $e^{-t/\theta}/\theta$ [Eq. (7.5)], is rather broad, as shown schematically in Figure 7.5a. The particle size distribution of latex products obtained from the two-continuous stirred tank reactor series is also quite broad [81]. As to the semibatch emulsion polymerization systems, they may result in very narrow or quite broad particle size distribution of latex products, depending on the recipe ingredients in the feed stream and the way it is added to the reactor. Latex products with different particle sizes and particle size distributions may exhibit quite different performance properties such as rheology, film formation, optical properties and perhaps storage stability (e.g., sedimentation of extremely large particles).

It was concluded that failure to address the above-mentioned problems and other significant factors that may appear for specific latex products means failure to develop an economically viable continuous emulsion polymerization process. A comprehensive understanding of the different characteristics between the batch or semibatch and continuous emulsion polymerization processes and how these differences affect product properties and reactor performance is a must for successful process development.

REFERENCES

1. G. W. Poehlein and D. J. Dougherty, *Rubber Chem. Technol.* **50**, 601 (1977).
2. B. Li and B. W. Brooks, *Polym. Int.* **29**, 41 (1992).
3. J. Snuparek, *Prog. Org. Coat.* **29**, 225 (1996).
4. J. Gao and A. Penlidis, *Prog. Polym. Sci.* **27**, 403 (2002).
5. C. S. Chern, in *Encyclopedia of Surface and Colloid Science*, A. Hubbard (Ed.), Marcel Dekker, New York, 2002, pp. 4220–4241.
6. M. Nomura, H. Tobita, and K. Suzuki, *Adv. Polym. Sci.* **175**, 1 (2005).
7. P. Fram, G. T. Stewart, and A. J. Szlochtum, *Ind. Eng. Chem.* **47**, 1000 (1955).
8. B. G. Elgood, E. V. Gulbekian, and D. Kinsler, *J. Polym. Sci. Part B* **2**, 257 (1964).
9. R. A. Wessling, *J. Appl. Polym. Sci.* **12**, 309 (1968).
10. H. Gerrens, *J. Polym. Sci. Part C* **27**, 77 (1969).
11. P. Bataille, B. T. Van, and Q. B. Pham, *J. Appl. Polym. Sci.* **22**, 3145 (1978).
12. J. Snuparek, *Angew. Makromol. Chem.* **25**, 113 (1972).
13. J. Snuparek and F. Krska, *J. Appl. Polym. Sci.* **20**, 1753 (1976).
14. J. Snuparek and F. Krska, *J. Appl. Polym. Sci.* **21**, 2253 (1977).
15. R. M. Fitch and R. C. Watson, *J. Colloid Interface Sci.* **68**, 14 (1979).
16. G. Lichti, R. G. Gilbert, and D. H. Napper, *J. Polym. Sci. Polym. Chem. Ed.* **21**, 269 (1983).

17. P. J. Feeney, D. H. Napper, and R. G. Gilbert, *Macromolecules* **17**, 2520 (1984).
18. R. W. Novak, *Adv. Org. Coat. Sci. Technol. Ser.* **10**, 54 (1988).
19. C. S. Chern and H. Hsu, *J. Appl. Polym. Sci.* **55**, 571 (1995).
20. N. Sutterlin, in *Polymer Colloids II*, R. M. Fitch (Ed.), Plenum Press, New York, 1980, p. 583.
21. K. Kronberg and P. Stenius, *J. Colloid Interface Sci.* **102**, 410 (1984).
22. R. J. Orr and L. Breitman, *Can. J. Chem.* **38**, 668 (1960).
23. L. M. Gugliotta, M. Arotcarena, J. R. Leiza, and J. M. Asua, *Polymer* **36**, 2019 (1995).
24. E. Unzueta, and J. Forcada, *Polymer* **36**, 4301 (1995).
25. C. S. Chern, H. Hsu, and F. Y. Lin, *J. Appl. Polym. Sci.* **60**, 1301 (1996).
26. C. S. Chern and C. H. Lin, *Polym. J.* **27**, 1094 (1995).
27. C. S. Chern and F. Y. Lin, *J. Macromol. Sci. Pure. Appl. Chem.* **A33**, 1077 (1996).
28. C. S. Chern and F. Y. Lin, *J. Appl. Polym. Sci.* **61**, 989 (1996).
29. C. S. Chern and C. H. Lin, *Polym. J.* **28**, 343 (1996).
30. C. S. Chern and Y. C. Chen, *Polym. J.* **28**, 627 (1996).
31. C. S. Chern and Y. C. Chen, *J. Macromol. Sci. Pure Appl. Chem.* **A35**, 965 (1998).
32. X. J. Xu and F. Chen, *Polymer* **45**, 4801 (2004).
33. C. S. Chern, C. H. Lin, and T. J. Chen, *Polym. J.* **29**, 249 (1997).
34. C. S. Chern, C. F. Huang, and C. L. Wang, *J. Disper. Sci. Technol.* **19**, 1 (1998).
35. S. Sajjadi and B. W. Brooks, *J. Appl. Polym. Sci.* **74**, 3094 (1999).
36. S. Sajjadi and B. W. Brooks, *J. Polym. Sci. Polym. Chem. Ed.* **38**, 528 (2000).
37. S. Sajjadi and B. W. Brooks, *Chem. Eng. Sci.* **55**, 4757 (2000).
38. C. S. Chern, T. J. Chen, S. Y. Wu, H. B. Chu, and C. F. Huang, *J. Macromol. Sci. Pure Appl. Chem.* **A34**, 1221 (1997).
39. M. Schneider, C. Graillat, A. Guyot, and T. F. McKenna, *J. Appl. Polym. Sci.* **84**, 1916 (2002).
40. S. Boutti, C. Graillat, and T. F. McKenna, *Macromol. Symp.* **206**, 383 (2004).
41. S. Sebenik and M. Krajnc, *Colloid Surface A: Physicochem. Eng. Aspects* **233**, 51 (2004).
42. C. S. Chern, *J. Appl. Polym. Sci.* **56**, 221 (1995).
43. C. S. Chern, *J. Appl. Polym. Sci.* **56**, 231 (1995).
44. Z. Wang, A. J. Paine, and A. Rudin, *J. Colloid Interface Sci.* **177**, 602 (1996).
45. C. S. Chern and Y. N. Kuo, *Chem. Eng. Sci.* **51**, 1079 (1996).
46. C. S. Chern and C. H. Lin, *Polym. Int.* **42**, 409 (1997).
47. S. Sajjadi, *Polymer* **44**, 223 (2003).
48. C. D. Immanuel, F. J. Doyle III, C. F. Cordeiro, and S. S. Sundaram, *AIChE J.* **49**, 1392 (2003).
49. S. Sajjadi and M. Yianneskis, *Macromol. Symp.* **206**, 201 (2004).
50. A. E. Hamielec and H. Tobita, in *Ulmann's Encyclopedia of Industrial Chemistry*, Vol. A21, VCH, Weinheim, Germany, 1992, p. 305.

51. G. W. Poehlein, in *Emulsion Polymerization and Emulsion Polymers*, P. A. Lovell and M. S. El-Aasser (Eds.), John Wiley & Sons, New York, 1997, p. 277.
52. D. B. Gershberg and J. E. Longfield, *Symposium On Polymerization Kinetics and Catalyst Systems*, Preprint No. 10, 45th AIChE Meeting, New York, 1961.
53. R. K. Greene, R. A. Gonzalez, and G. W. Poehlein, in *Emulsion Polymerization*, I. Piirma and J. L. Gardon (Eds.), ACS Symposium Series 24, American Chemical Society, Washington, D.C. 1976, p. 341.
54. M. Nomura and M. Harada, in *Emulsion Polymers and Emulsion Polymerization*, D. R. Bassett and A. E. Hamielec (Eds.), ACS Symposium Series 165, American Chemical Society, Washington, D.C. 1981, p. 121.
55. M. Nomura, S. Sasaki, W. Xue, and K. Fujita, *J. Appl. Polym. Sci.* **86**, 2748 (2002).
56. C. Kiparissides, J. F. MacGregor, and A. E. Hamielec, *J. Appl. Polym. Sci.* **23**, 401 (1979).
57. J. B. Rawlings and W. H. Ray, *Polym. Eng. Sci.* **28**, 237 (1988).
58. J. B. Rawlings and W. H. Ray, *Polym. Eng. Sci.* **28**, 257 (1988).
59. K. Tauer and I. Muller, *DECHEMA Monogr.* **131**, 95 (1995).
60. H. Gerrens, K. Kuchner, and G. Ley, *Ind. Tech.* **43**, 693 (1971).
61. G. Ley and H. Gerrens, *Makromol. Chem.* **175**, 563 (1974).
62. A. R. Berrens, *J. Appl. Polym. Sci.* **18**, 2379 (1974). ???
63. S. Omi, T. Ueda and H. Kubota, *J. Chem. Eng. Jpn.* **2**, 193 (1969).
64. M. Nomura, H. Kojima, M. Harada, W. Eguchi, and S. Nagata, *J. Appl. Polym. Sci.* **15**, 675 (1971).
65. A. W. DeGraff and G. W. Poehlein, *J. Polym. Sci. Part A-2* **9**, 1955 (1971).
66. G. W. Poehlein, W. D. Dubner, and H. C. Lee, *Br. Polym. J.* **14**, 143 (1982).
67. G. W. Poehlein, H. C. Lee, and N. Stubicar, in *Polymer Reaction Engineering*, K. H. Reichert and W. Geiseler (Eds.), Hanser Publishers, Munich, 1983, p. 113.
68. H. C. Lee and G. W. Poehlein, *J. Dispersion Sci. Technol.* **5**, 247 (1984).
69. G. W. Poehlein, H. C. Lee, and N. Stubicar, *J. Polym. Sci. Polym. Symp.* **72**, 207 (1985).
70. H. C. Lee, Emulsion Polymerization in a Seed-Fed Continuous Stirred-Tank Reactor, Ph.D. Dissertation, School of Chemical Engineering, Georgia Institute of Technology, Atlanta, GA, 1985.
71. H. C. Lee and G. W. Poehlein, *Polym. Process Eng.* **5**, 37 (1987).
72. J. Ugelstad, P. C. Mork, and J. O. Aasen, *J. Polym. Sci. A-1* **5**, 2281 (1967).
73. W. H. Stockmayer, *J. Polym. Sci.* **24**, 314 (1957).
74. J. L. O'Toole, *J. Appl. Polym. Sci.* **9**, 1291 (1965).
75. S. Katz and G. M. Saidel, *Polymer Preprints* **7**, 737 (1966).
76. S. Katz and G. M. Saidel, *AIChE J.* **13**, 319 (1967).
77. S. Katz, R. Shinnar, and G. M. Saidel, *Adv. Chem. Ser.* **91**, 145 (1969).
78. G. M. Saidel and S. Katz, *J. Polym. Sci. Part C* **27**, 149 (1969).
79. G. Lichti, R. G. Gilbert, and D. H. Napper, *J. Polym. Sci. Polym. Chem. Ed.* **18**, 1297 (1980).

80. R. G. Gilbert, *Advances in Emulsion Polymerization and Latex Technology*, 12th Annual Short Course Notes, Lecture 12, Emulsion Polymers Institute, Lehigh University, Bethlehem, PA, 1982.

81. G. Poehlein, in *Polymerization Reactors and Processes*, J. N. Henderson and T. C. Bouton (Eds.), ACS Symposium Series 104, American Chemical Society, Washington, D.C. 1979, p. 1.

8

EMULSION POLYMERIZATIONS IN NONUNIFORM LATEX PARTICLES

8.1 ORIGIN OF NONUNIFORM LATEX PARTICLES

Multiphase polymer particles prepared by emulsion polymerizations find a number of important commercial applications such as elastomers, coatings, adhesives, and impact resistant thermoplastics. Latex products, which exhibit nonuniform particle morphology, are produced when two or more monomers react with one another such that separate polymer phases form during emulsion polymerization. The incompatibility of different polymers or the sequence and location of the formation of polymers can result in separate polymer phases.

Multistage emulsion polymerizations have been widely used to design and synthesize latex particles with a variety of morphologies. As soon as an appreciable amount of the postformed polymer is generated in the seed particles, a two-phase structure will normally be observed within the latex particles because of the relatively incompatible nature of most polymer pairs (Figure 8.1). The resultant emulsion polymers with different particle morphologies generally possess different mechanical and physical properties. Thus, design and control of latex particle morphology is crucial in order to fulfill some end-use requirements. Published studies on the morphology of nonuniform latex particles are rather limited. Very often, contradictory results exist in the literature because latex particle morphology is a very complicated function of a number of physical parameters (e.g., the incompatibility of polymer pairs, distribution of free radicals and monomers in the polymerizing latex particles, degree of grafting at the interfacial layer of polymer pairs, polymer molecular

Principles and Applications of Emulsion Polymerization, by Chorng-Shyan Chern

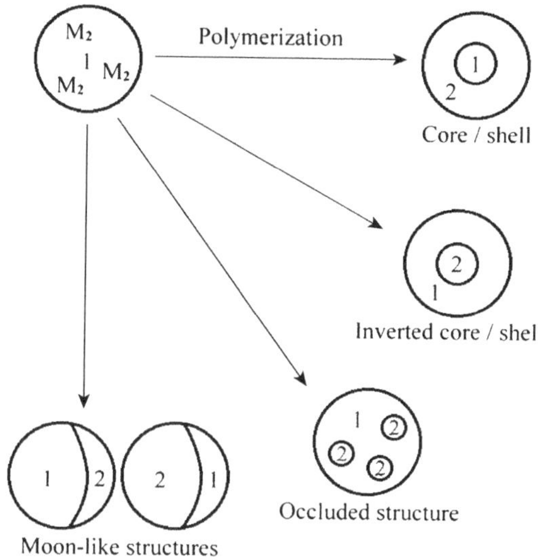

Figure 8.1. Some possible latex particle morphologies obtained from the seeded emulsion polymerization of monomer 2 in the presence of seed particles of polymer 1.

weight, glass transition temperatures of polymer pairs, etc.) and reaction conditions (e.g., the recipes, sequence and location of the formation of polymer chains, polymerization temperature, etc.).

This chapter serves as an introduction to the origin of nonuniform latex particles. First, a brief discussion of the seeded emulsion polymerization technique that has been widely used to prepare composite polymer particles with a variety of morphological structures is given. This is followed by the illustration of the effects of important factors such as initiators, monomer addition methods, polymer molecular weight, volume ratio of the second-stage monomer to the seed polymer, and polymerization temperature that affect the morphological structures of latex particles. The development of morphological structures of nonuniform latex particles will then be covered at the end of this chapter.

8.2 SEEDED EMULSION POLYMERIZATIONS

If a seed latex is swollen with monomer, which is different from the seed polymer, phase separation will usually take place in subsequent emulsion polymerization (termed the seeded emulsion polymerization). This will then lead to a variety of morphological structures for the resultant latex particles [1–10]. By rule of thumb, the greater the incompatibility of polymer pairs, the greater the extent of phase separation in the polymerizing latex particles. The polarity of polymer is a measure of the compatibility of polymer pairs. The larger the difference in the polarity of polymer, the greater the incompatibility of polymer pairs. The polarities of some representative emulsion polymers in the decreasing order are polyvinyl chloride (9.75, 9.59) > polyvinyl acetate (9.65, 9.46) > polymethyl methacrylate (9.67, 9.24) > poly(*n*-butyl acrylate) (9.03, 8.92) > polystyrene (8.97, 9.01) > polybutadiene (8.17, 9.12). The first numeric value in the parentheses represents the average solubility parameter obtained from all the data available in the *Polymer Handbook* [11], whereas the second number is the calculated value according to the group contribution method of Small [11, 12]. As discussed in Chapter 2, Section 2.2.3, the solubility parameter measures the intermolecular interactions (e.g., hydrogen bonding, dipole–dipole, and van der Waals interactions) of materials. The larger the difference in the solubility parameter, the stronger the incompatibility between a pair of polymers. For example, the degree of phase separation between polyvinyl acetate and poly(*n*-butyl acrylate) in the two-phase emulsion polymer particles is larger; therefore, more heterogeneous latex particles are produced during that seeded emulsion polymerization as compared to the polystyrene–poly(*n*-butyl acrylate) pair. It should be noted that even complete compatibility between two polymers does not guarantee the formation of a uniform latex particle structure. The latex particle morphology may be dependent on other factors such as the distribution of free radicals and monomers in the polymerizing particles, methods of monomer addition, and so on.

If the seed latex particles can barely be swollen by the second-stage monomer and a water-soluble initiator is used, then the subsequent seeded emulsion polymerization will be localized near the particle surface layer. Thus, the postformed polymer tends to form a surface layer around the seed latex particle. An example of this kind of morphological structure of latex particles is the seeded emulsion polymerization of methyl methacrylate in the presence of a polyvinylidene chloride seed latex. On the other hand, free radical polymerization can take place inside the seed latex particles. In this manner, various morphological structures of latex particles such as the perfect core/shell, inverted core/shell, dumbbell-shaped, and occluded structures can be achieved, depending on various physical parameters and polymerization conditions.

8.3 FACTORS AFFECTING PARTICLE MORPHOLOGY

8.3.1 Effect of Initiators

Emulsion polymerization using a water-soluble initiator generates oligomeric radicals in the continuous aqueous phase, which are hydrophilic and often anionic (e.g., the sulfate end-group derived from persulfate initiator). When these oligomeric radicals become hydrophobic enough, they can enter the monomer-swollen polymer particles. Furthermore, the negatively charged sulfate end group of the oligomeric radical tends to remain on the latex particle surface and thereby constrain the movement of the growing free radical into the interior of the particle [13]. Such a nonuniform distribution of free radicals in the reaction loci during seeded emulsion polymerization will have a significant influence on the morphological structures of latex particles [6]. This is because most free radical polymerization takes place near the latex particle surface layer, and the resultant polymer tends to encapsulate the seed particle and form a shell. Nevertheless, the formation of a perfect core/shell structure is not ensured without considering other important factors such as the polarity of the polymers and polymerization conditions.

When an oil-soluble initiator is used, free radicals distribute themselves, in a stochastic sense, more uniformly in the latex particles. The subsequent emulsion polymerization then yields a morphological structure in which the post-formed polymer appears not only near the latex particle surface layer but also inside the composite particle, as observed by Merkel et al. [14].

Cho and Lee [6] used three different initiators, potassium persulfate, 2,2′-azobisisobutyronitrile, and 4,4′-azobis(4-cyanovaleric acid) (water-soluble, but less hydrophilic than potassium persulfate) to investigate their effects on the emulsion polymerization of styrene in the presence of polymethyl methacrylate seed latex particles. Inverted core/shell latex particles were observed when 2,2′-azobisisobutyronitrile or 4,4′-azobis(4-cyanovaleric acid) was used to initiate free radical polymerization. The use of potassium persulfate resulted in various morphological structures of latex particles, which were largely determined by the initiator concentration and polymerization temperature.

8.3.2 Effect of Monomer Addition Methods

Min et al. [5] investigated the two-stage polystyrene (second stage)/poly(*n*-butyl acrylate) (first stage) emulsion polymerization. Sodium dodecyl benzyl sulfonate and potassium persulfate were used as the surfactant and initiator, respectively. A thin-layer chromatography/flame ionization detector was employed to determine the degree of grafting, which is dependent on the method of monomer addition. The degree of grafting was the greatest for the batch emulsion polymerization process with little time for seed particle swelling, less for the preswollen batch process, and least for the semibatch process. When the extent of grafting was relatively low, the postformed polystyrene separated from the preformed poly(*n*-butyl acrylate) to form the dumbbell-shaped latex particles upon aging for one year. On the other hand, the latex particles remain spherical when the degree of grafting was high enough. Graft polystyrene generated during the second stage of emulsion polymerization tends to remain in the interfacial region of polystyrene–poly(*n*-butyl acrylate) particles, and it helps to stabilize the morphological structure of latex particles. The influence of grafting efficiency on the morphological structures of poly(methyl methacrylate)/polybutadiene and poly(methyl methacrylate)/styrene–butadiene copolymer latex particles prepared by the seeded emulsion polymerization technique was studied by Merkel et al. [14].

Continuous reactors with seed latex particles in the feed stream could be an interesting polymerization system for morphological studies. The broad residence time distribution of the polymerizing latex particles associated with such a reactor configuration results in a broad particle size distribution of the effluent product. By changing the particle size distribution (monodisperse or polydisperse) of seed latex particles and operation conditions (mean residence time, monomer addition method, etc.) simultaneously, one can essentially obtain a variety of morphological structures of latex particles.

8.3.3 Effect of Polymer Molecular Weight

Lee [3] used a chain transfer agent to study the effect of polymer molecular weight on the morphological structures of the emulsion polymerization of styrene in the presence of styrene–butadiene rubber seed latex particles. Sodium dodecyldiphenyl oxide sulfonate, sodium persulfate, and bromoform and carbon tetrachloride were used as the surfactant, initiator, and chain transfer agents, respectively. At intermediate levels of chain transfer agent, the postformed polystyrene chains separate as microdomains within the styrene–butadiene rubber seed latex particles. This is followed by a phase separation process; as a result, the formation of a continuous polymer phase as the amount of the second-stage polystyrene is increased. At high levels of chain transfer agent, however, polystyrene chains with relatively low molecular weight become completely separated from the styrene–butadiene rubber seed copolymer. This then resulted in a hemispherical morphology when the quantity of polystyrene was equal to the styrene–butadiene rubber seed copolymer. By further increasing the second-stage polymer, an asymmetric encapsulation structure of the resultant composite latex particles was obtained.

Cho and Lee [6] varied the viscosity of the reaction loci and investigated its effect on the morphological structure of latex particles. The reaction system chosen for study was the emulsion polymerization of styrene in the presence of polymethyl methacrylate seed latex particles. Viscosity was controlled by the ratio of styrene to polymethyl methacrylate and the addition of a chain transfer agent or a solvent that is common to both polystyrene and polymethyl methacrylate. They also demonstrated that the latex particle morphology was affected by polymer molecular weight. The possibility of branching and crosslinking reactions increases dramatically during the latter stage of emulsion polymerization or in monomer-starved polymer reactions. Branching and crosslinking reactions will significantly increase the polymer molecular weight. The postformed polymer may be trapped inside the crosslinked seed polymer particles to form two interpenetrating phases surrounded by a shell rich in the postformed polymer.

Another scenario about the impact of crosslinked seed particles is briefly described as follows. When a crosslinked seed latex is swollen by second-stage monomers, the crosslinked network structure expands due to the solvency power of the newly incorporated monomers. When those monomers are converted to incompatible postformed polymer, however, the crosslinked network structure of preformed polymer will try to contract to a more equilibrium configuration and the postformed polymer may be squeezed out to form non-spherical latex particles. This type of latex products exhibits a unique rheological behavior (see Chapter 9, Section 9.2).

8.3.4 Effect of Volume Fractions of Polymer Pairs

The seed polymer particle tends to precipitate inside the final latex particle and sometimes escapes from the particle when the volume ratio of the second-stage monomer to the seed polymer is very high. As another extreme, very often the postformed polymer cannot encapsulate the seed polymer particle completely and, therefore, the formation of a perfect core/shell latex particle becomes almost impossible if the volume ratio of the second-stage monomer to the seed polymer is very small. Taking the two-stage poly(methyl methacrylate) (second stage)/polystyrene (first stage) emulsion polymerization system as an example, the calculated ratios of the shell thickness of the resultant latex particle to the core seed particle diameter, which is independent of the seed particle size, in decreasing order are 1.22 (10/1) > 0.82 (5/1) > 0.44 (2/1) > 0.26 (1/1) > 0.14 (0.5/1) > 0.032 (0.1/1). The numeric value in the parentheses is the corresponding volume ratio of the second-stage monomer to the first-stage seed polymer. The assumption involved in these calculations is that nucleation of a second crop of particle nuclei and flocculation of latex particles do not take place during the hypothetical seeded emulsion polymerization and the resultant latex particles form a perfect core/shell structure.

8.3.5 Effect of Polymerization Temperature

Polymerization temperature can affect the mobility of both monomer and polymer species and, hence, the rate of phase separation of polymer pairs, which will consequently contribute to the determination of the latex particle morphology.

8.4 MORPHOLOGY DEVELOPMENT IN LATEX PARTICLES

8.4.1 Thermodynamic Considerations

Sundberg et al. [15] and Chen et al. [16] adopted thermodynamic considerations to minimize the total Gibbs free energy changes associated with the latex particle morphologies. This approach shows the role of the interfacial tension in controlling the latex particle morphology, and some success in predicting the particle morphology has been achieved. It should be noted that such a thermodynamic analysis only predicts the ultimate latex particle morphology when the aging time approaches infinity; unfortunately, this is generally not the case because other physicochemical parameters and polymerization conditions may also come into play in determining the particle morphology. Durant and Sundberg [17] developed an algorithm to predict the thermodynamic equilibrium morphology of latex particles as a function of monomer conversion. It was pointed out that the predicted latex particle morphology should match that observed experimentally for the interfacial free energy surface with steep contours adjacent to the minimum point. On the other hand, there are a number of possible latex particle morphologies, which possess comparable interfacial free energies, in the emulsion polymerization system when the minimal point is located within a rather flat region on the interfacial free energy surface. Several case studies dealing with composite particles comprising polystyrene and polymethyl methacrylate and two very different surfactants were carried out.

Gonzalez and Asua [18] developed an interesting mechanistic model for the migration of clusters during the latex particle morphology development in emulsion polymerization. It was postulated that the driving force for the migration of clusters is the balance between the van der Waals force and viscous force. Several equilibrium latex particle morphologies (e.g., core/shell, inverted core/shell, and occluded structure) predicted by this dynamic model were illustrated. Later, the authors considered composite particles to be a biphasic system comprising clusters of polymer 1 dispersed in a matrix of polymer 2 [19]. Both the polymerization and migration of clusters were incorporated into the framework of the model. The free radical polymerization of monomer 1 takes place in both the matrix of polymer 2 and in clusters of polymer 1. Furthermore, polymer 1 produced in the matrix of polymer 2 can diffuse readily into clusters of polymer 1. Migration of clusters is directed toward the equilibrium latex particle morphology in order to minimize the interfacial free energy of the seeded emulsion polymerization system. The driving forces for the migration of clusters include van der Waals forces between the clusters and the continuous aqueous phase and those between the clusters themselves. In addition, the effect of the viscosity of the matrix of polymer 2 on the migration of clusters was also taken into consideration. They then developed another model that takes into account the phase separation to induce the nucleation of clusters, polymer reactions, diffusion of polymer chains, and migration of clusters [20]. The model was used to simulate the emulsion polymerization of methyl methacrylate in the presence of polystyrene seed particles. A relatively reasonable agreement between the experimental results and model predictions was achieved. Furthermore, both the initial volume of clusters and the rate coefficient for the nucleation of clusters have an insignificant influence on the equilibrium latex particle morphology.

8.4.2 Nonequilibrium Morphology Development

Aerdts et al. [21] first synthesized polybutadiene seed latex particles (90 nm in diameter). This was followed by the semibatch emulsion copolymerization of styrene and methyl methacrylate in the presence of the rubber seed latex particles. Finally, glycidyl methacrylate was grafted onto the composite latex particles. The emulsion polymerizations were stabilized by sodium dihexyl sulfosuccinate and initiated by potassium persulfate. The functionalized composite polymer particles, which served as an impact modifier, were blended with polyamide-6, and these particles were dispersed well in the matrix of polyamide-6.

Stubbs et al. [22] also studied the nonequilibrium development of latex particle morphologies in seeded emulsion polymerizations. They interpreted the effects of various experimental variables on the final latex particle morphology through an assessment of probabilities for diffusion and reaction of both polymeric radicals and monomer molecules within the particle. It was concluded that a uniform monomer concentration profile across the latex particle can be achieved without any problem, even during the slow monomer feed period or for the glassy seed polymer particle. By contrast, polymeric radicals are most likely restricted near the latex particle surface layer when the flux of free radicals is high enough and the rate of monomer feed is slow enough for the glassy seed polymer particle. Nevertheless, this might not be the case for the soft seed polymer particle. Thus, the free radical polymerization in the presence of glassy seed polymer particles is characterized by the "monomer-starved feed" process, whereas the "monomer-starved feed" process is not experienced in the polymerization system using soft seed polymer particles. Under these circumstances, the probability for the formation of nonequilibrium latex particle morphologies is higher in the former case.

Karlsson et al. [23] prepared and characterized a series of latex products with heterogeneous particle structures via seeded emulsion polymerizations. The weight percentage of the relatively hydrophobic seed latex particles (poly(styrene-*co*-*n*-butyl acrylate) with glass transition temperatures ranging from 20 °C to 100 °C) is kept constant at 40%. The relatively hydrophilic second-stage polymer comprises monomeric units of methyl methacrylate, *n*-butyl acrylate, and methacrylic acid with the glass transition temperature equal to 20 °C. Calculated rates of diffusion of the propagating species during emulsion polymerization were correlated well with the observed morphological evolution and the fraction of interface in the composite polymer particles. Stubbs and Sundberg [24] studied the effect of the second-stage initiator end-groups on the development of latex particle morphology in the semibatch emulsion polymerization of styrene in the presence of poly(methyl acrylate-*co*-methyl methacrylate) seed latex particles. Anionic potassium persulfate and nonionic 2,2′-azobis(2-methyl-*N*-(2-hydroxylethyl) propionamide) were the initiators chosen for this study. This work illustrates that there exist intermediate diffusion rates of polymeric radicals that can highlight the effect of electrostatically charged and uncharged initiators on the development of latex particle morphology. Diffusion rates of polymeric radicals that are either much faster or much slower result in the same morphological structure of the latex particles. The chain anchoring effect arising from initiator end-groups alone is expected to restrict the penetration of polymeric radicals toward the interior region of the latex particle, but it was seldom the predominant factor in the evolution of morphological structures. Other parameters such as the glass transition temperature of seed polymer, polymerization temperature, initiator concentration, and monomer feed rate have more direct control on the diffusivity of polymeric radicals inside the latex particle. In addition, chain transfer reactions of polymeric radicals to monomer and chain transfer agent (if present) promote penetration of free radicals within the latex particle. Significant penetration of free radicals in the latex particle often results in structures comprising a number of occlusions, even when an apparent shell is observed.

Stubbs and Sundberg [25] then carried out a series of experiments to investigate the effect of *n*-dodecyl mercaptan (a chain transfer agent) on the development of morphological structures in latex particles. The results show that the addition of *n*-dodecyl mercaptan enhances the extent to which the second stage polymer domains form within the interior region of the seed latex particle, but this phenomenon is only expected under specific polymerization conditions. In addition, the greatly reduced molecular weight of the second-stage polymer due to the addition of chain transfer agent did not significantly increase the extent of phase separation between polymer pairs and affect the resultant latex particle morphology. The overall effect of the chain transfer agent on the morphological structures of latex particles was rather limited. On the other hand, crosslinking reactions occurring in the second stage of polymerization generate branched and crosslinked polymer chains that diffuse slowly within the latex particle in comparison with their linear counterparts. Thus, a perfect core/shell structure is expected to form in this seeded emulsion polymerization. However, crosslinking reactions taking place in the emulsion copolymerization of styrene and a few percent divinyl monomer in the presence of a noncrosslinked, polar methacrylic seed polymer exhibited very little effect on the latex particle morphology [26]. This is because the probability for polymeric radicals to develop a branched chain before penetrating a significant distance into the interior region of the latex particle is very low. On the other hand, crosslinking reactions in the seeded emulsion polymerization retard phase separation of the second-stage polymer from the seed polymer and hinder rearrangement of phase-separated microdomains as well. With normal crosslinking agent concentrations (a few percent or lower), this effect is also insignificant in the development of the latex particle morphology. However, at intermediate crosslinking agent concentrations, smaller sizes of phase-separated microdomains are achieved in seeded emulsion polymerizations. At high crosslinking agent concentrations, phase separation of polymer pairs becomes severely retarded.

Ferguson et al. [27] explored how to prepare latex particles with a core/shell (polystyrene/polyvinyl acetate) structure. A variety of initiators including potassium persulfate, ammonium persulfate, 2,2′-azobisisobutyronitrile, and benzoyl peroxide were used as the initiator. To minimize the secondary nucleation of particle embryos, polyvinyl acetate was chosen as the seed latex particles and seeded emulsion polymerizations of the second-stage monomer (styrene) were carried out. The objective of this work was to prepare latex particles with an ideal core/shell structure via the phase inversion of the second-stage polystyrene formed during the seeded emulsion polymerization and the first-stage seed polymer (polyvinyl acetate), as would be expected by thermodynamic predictions. It was shown that evolution of the target latex particle morphology is promoted by fast diffusion of polyvinyl acetate chains toward the particle surface layer during the emulsion polymerization of styrene in the presence of polyvinyl acetate seed particles. A number of methods (e.g., reducing the molecular weight and degree of branching of polyvinyl acetate, minimizing the grafting reaction of polystyrene radicals onto polyvinyl acetate chains, and increasing the hydrophilicity of polyvinyl acetate by incorporating a small amount of hydrophilic comonomers such as sodium salt of vinylsulfonic acid) to achieve this goal were proposed. Zhao et al. [28] investigated the evolution of the latex particle morphology in the emulsion polymerization of *n*-butyl acrylate in the presence of polyvinyl acetate seed particles stabilized by mixed anionic and nonionic surfactants and initiated by potassium persulfate at 70 °C. The emulsion polymerization system was operated under the monomer-starved condition. The inverted core/shell [poly(*n*-butyl acrylate)/polyvinyl acetate] structure is the thermodynamically preferred latex particle morphology. Nevertheless, multiparticle morphologies were observed, which was attributed to the restricted polymer chain mobility closely related to the high viscosity inside the polymerizing latex particles. It was postulated that poly(*n*-butyl acrylate) first forms around polyvinyl acetate seed particles and then migrates toward the core region of the seed latex particles.

Pan et al. [29] first prepared crosslinked poly[(*n*-butyl acrylate)-*co*-2-ethylhexyl acrylate]/poly(methyl methacrylate-*co*-styrene) (core/shell) latex particles by the multistage emulsion polymerization. The core/shell latex particles were then used as the seed particles in the emulsion polymerization of vinyl chloride. Sodium dodecyl sulfate and potassium persulfate were used as the surfactant and initiator, respectively. The efficiency of grafting polyvinyl chloride radicals onto the seed copolymer chains increased with increasing amount of seed latex particles. The polymer particle morphology was transformed from the perfect core/shell structure into irregular sandwich-like structures when the weight percentage of the monomeric units of styrene in the shell was greater than 70%. Furthermore, the composite particles prepared by the emulsion polymerization of vinyl chloride in the presence of these seed latex particles exhibit a three-layered core/shell structure. The compatibility between polyvinyl chloride and these composite seed latex particles was good, and these seed particles were dispersed uniformly in the polyvinyl chloride matrix.

8.4.3 Techniques for Characterization of Particle Morphology

Synthetic latex particles comprising two or more individual polymers have found versatile applications in areas such as coatings, adhesives, binders, and impact modifiers for brittle plastics. A variety of morphological structures (Figure 8.1) of composite latex particles have been observed. It is crucial, however, to establish the latex particle structure–property relationships in order to design latex products with satisfactory performance properties.

In the past 50 years, several analytical techniques were developed to characterize morphological structures of composite latex particles [30]. These characterization techniques include scanning electron microscopy (SEM), transmission electron microscopy (TEM), atomic force microscopy (AFM), solid-state nuclear magnetic resonance (NMR), differential scanning calorimeter (DSC), and minimum film formation temperature (MFFT). Among these techniques, electron microscopy is perhaps the most useful tool for studying the latex particle morphology. This is especially true when the latex particles are sectioned in a microtome prior to viewing in the TEM.

AFM is usually used to examine the whole latex particles. High-energy neutrons [small-angle neutron scattering (SANS) or X rays (small-angle X-ray scattering (SAXS)] can be scattered by the multiple polymer phases within composite particles; therefore, these analysis methods provide invaluable information on latex particle morphology. Solid-state NMR has also been used to characterize nonuniform latex particles [31–33]. Thermal analysis techniques such as DSC are used to study the morphological structures of composite latex particles [34, 35]. Finally, MFFT tests can be performed to determine a narrow temperature range over which the dried latex particles are transformed from a powder to an integral polymer film. This approach more or less reflects the morphological structures of composite latex particles.

Recently, Stubbs and Sundberg [36] completed an interlaboratory study (i.e., a round-robin study) to determine the morphological structures of composite latex particles obtained from a particular emulsion copolymerization of styrene and *n*-butyl acrylate ($T_g = 50\,°C$) in the presence of polymethyl methacrylate seed particles (71 nm in diameter). Sodium dodecyl sulfate and potassium persulfate were used as the surfactant and initiator, respectively. The polymerization temperature is 70 °C. Under this circumstance, the seed polymer would be glassy, whereas the second-stage copolymer would be soft during the reaction. Six independent organizations participated in this interesting study. The complementary characterization techniques used included SEM, TEM, AFM, solid-state NMR, DSC, MFFT, capillary hydrodynamic fractionation (CHDF) chromatography, and dynamic light scattering (DLS). The results indicate that a complete determination of the latex particle morphology requires characterization of three major aspects including (a) the overall particle shape, (b) the composition of the particle surface, and (c) the internal particle structure. For the same characterization technique, the agreement between different laboratories was satisfactory, especially for the nonmicroscopy techniques such as DSC and surfactant titration that is used to characterize the particle surface properties. The microscopy techniques, especially TEM, show the most variation between different laboratories due to the varying sample preparation methods and instrument operation skill.

Recently, Stubbs and Sundberg [37] used the modulated temperature DSC to investigate the extent of phase separation of polymer pairs during polymerization in the semibatch seeded emulsion polymerization system. The two polymerization systems chosen for this study included (a) poly(methyl methacrylate-*co*-methyl acrylate)/polystyrene and (b) poly(styrene-*co*-*n*-butyl acrylate)/polymethyl methacrylate (postformed polymer/preformed polymer). The emulsion polymerization systems were stabilized by sodium dodecyl sulfate and initiated by potassium persulfate. The two polymer pairs (i.e., the seed emulsion polymer and the second-stage polymer) exhibited very different polarity characteristics, and these recipes in combination with different monomer feed rates were used to provide various degrees of dynamic polymer phase separation for this fundamental latex particle morphology study. This work clearly demonstrates that modulated temperature DSC data can be effectively used to follow the polymer phase separation process with the progress of monomer feed. This approach allows the quantitative estimation of the relative amounts of different polymer phases and their compositions during polymerization. As expected, the polymer phase separation process is very often far from complete at the end of the polymerization.

Araki et al. [38] studied the feasibility of applying resonant soft X-ray scattering to chemically heterogeneous soft condensed matter nanomaterials. Two structured styrene-acrylic emulsion polymer particles with average particle diameters close to 230 nm were examined. This technique can be used to obtain the effective radii corresponding to the two polymer phases within the latex particles, and it can serve as a powerful complementary tool to neutron and hard X-ray scattering techniques for the characterization of structured soft condensed matter nanomaterials.

8.5 POLYMERIZATION KINETICS IN NONUNIFORM LATEX PARTICLES

Development of morphological structures of nonuniform latex particles during polymerization has attracted much attention of polymer scientists associated with coatings, adhesives, binders, impact modifiers, and so on. An excellent review of this research area specifically dealing with the latex particle morphologies at thermodynamic equilibrium and those in nonequilibrium state is provided in reference 39. The evolution of latex particle morphologies in the nonequilibrium state is primarily controlled by the rate of diffusion of polymeric radicals within the polymerizing particles and the rate of polymer phase separation and, ultimately, phase rearrangement. In turn, the rate of diffusion of polymeric radicals is dependent on the rate of growth of active polymer chains within the nonuniform latex particles. The absorption of oligomeric radicals by the latex particles, the bimolecular termination of free radicals and chain transfer of free radicals to monomer, polymer, or chain transfer agent (if present), transport of free radicals between the polymer phases, desorption of free radicals out of the particles, partitioning of monomer between the two polymer phases, and reaction temperature need to be taken into consideration in order to gain a better understanding of the kinetics of two-phase emulsion polymerizations.

8.5.1 Pioneering Studies

Rosen [40] is a pioneer in the field of two-phase emulsion polymerization mechanisms and kinetics. He studied the influence of the polymer phase separation on the efficiency of grafting reactions and demonstrated the physical limits of these grafting reactions. This was followed by the work of Chiu [41] with an attempt to develop a mechanistic model that can be used to predict the experimental kinetic data obtained from two-phase emulsion polymerizations that was not successful.

Varshney [42] experimentally studied two-phase emulsion polymerization systems of polystyrene/polymethyl methacrylate and polymethyl methacrylate/polystyrene (postformed polymer/preformed polymer) and showed that a single-phase model is incapable of adequately predicting the polymerization kinetic data. He also reported that the distribution coefficient for the second-stage monomer in the two-polymer phases is close to unity. Based on the Smith–Ewart theory [43], Nelson [44] and Nelson and Sundberg [45] then developed kinetic models for two-phase emulsion polymerization systems with core-shell, inverted core-shell, and occluded particle morphologies. They assumed that free radical polymerization takes place in both polymer phases of the latex particles. A free radical population balance in combination with a modified Stockmayer formula was established to predict the average number of free radicals in each polymer phase. The free radical population balance relationship was obtained from a consideration of the following events: generation of free radicals in the continuous aqueous phase, transport of free radicals between polymer phases, and bimolecular termination of free radicals in each polymer phase. Transport of free radicals out of the latex particles was considered negligible in their kinetic model. The resultant average number of free radicals in each polymer phase was used to calculate the rate of polymerization for each polymer phase. A total polymerization rate was then obtained by summing the reaction rates for each polymer phase. The gel effect (i.e., autoacceleration in the polymerization rate with monomer conversion) was taken into account in the kinetic expressions by a significant reduction in the termination rate constant with increasing monomer conversion. This effect was quantitatively correlated to changes in the free volume of the reaction system with the progress of polymerization. The polymerization system that falls short of complete monomer conversion (i.e., limiting conversion) was also considered in a similar manner. This unique phenomenon was attributed to the decreased diffusion coefficient for monomer (i.e., propagation rate constant) at high monomer conversion. The mechanistic model was then employed to simulate monomer conversion versus time data for different two-phase emulsion polymerization systems, and it could predict the trends of the polymerization rate data. One important conclusion obtained from this work was that evolution of morphological structures and reaction kinetics are mutually dependent processes in two-phase emulsion polymerization.

This elegant approach requires an overall mass transfer coefficient for the exchange of free radicals between the two polymer phases in order to carry out computer simulations [44, 45]. This parameter was estimated by applying the Sherwood number relationship to two-phase emulsion polymerization systems. The core-phase was regarded as the sphere and the shell-phase as the stagnant fluid in the mass transfer treatment. The thickness of the postformed polymer shell, which would increase from zero to a finite value with increasing monomer conversion, was chosen as the characteristic length in the Sherwood number relationship. Only monomeric radicals were considered to be capable of exchanging between polymer phases. Hence, the overall mass transfer coefficient was corrected by multiplying by a factor of $k_{tr,m}/k_p$, which is valid when chain transfer to monomer controls the polymer molecular weight. The parameters $k_{tr,m}$ and k_p are the rate constants for the chain transfer reaction of a propagating radical with monomer and propagation reaction, respectively. This treatment results in a mathematical problem at the very beginning of polymerization in that the overall mass transfer coefficient tends to be unreasonably large at very small thickness of the shell-phase. Cornish [46] pointed out that in heat transfer for ratios of radii (shell : core) of 2, 5, 10, and 50, the corresponding Nusselt numbers are 4, 2.5, 2.22, and 2.04, respectively. The work of Cornish also introduces uncertainties into the approach used by Sundberg et al. in calculations of the overall mass transfer coefficient. In addition, transport of free radicals out of the latex particles was ignored in Sundberg's model. This assumption is only true for the conditions of large latex particles and/or small bimolecular termination rates in polymer phases.

8.5.2 Effect of Distribution of Free Radicals in Nonuniform Latex Particles

The significance of work of Nelson and Sundberg [44, 45] is to illustrate the viability of modeling two-phase emulsion polymerization kinetics by using concepts very similar to those used for single-phase emulsion polymerization systems. Nevertheless, they did not take into account the fact that each free radical that enters the latex particle from the continuous aqueous phase has a hydrophilic sulfate end-group and will preferentially remain in the particle surface layer. This distinctive characteristic of conventional emulsion polymerization systems would cause nonuniform distribution of free radicals in polymerizing latex particles and might significantly affect the two-phase emulsion polymerization kinetics. This is because the hydrophilic sulfate end-group of a polymeric radical that is anchored onto the latex particle surface layer has a significant influence on the particle morphology development. In turn, this free radical anchoring effect is expected to play an important role in the polymerization kinetics of two-phase latex particles.

The sulfate end-group tends to constrain the movement of free radicals into the interior of the latex particle. It is this constrained end-group location that is responsible for the nonuniform distribution of free radicals within the latex particle. A very simple model using a Monte Carlo technique was developed for computing the distribution of free radicals in latex particles with the constraint that the hydrophilic end-groups of these growing polymer chains remain on the particle surface [13]. Computer simulations with this model reflect that the concentration of free radicals will be statistically greater near the latex particle surface. For example, the calculated free radical distribution function for a latex particle with a diameter of 400 nm, $\rho(z)$, is a decreasing exponential function, as shown in Eq. (8.1):

$$\ln \rho(z) = -1.4259 - 4.3787z \tag{8.1}$$

where $z = 0$ and $z = 1$ represent the position at the particle surface and at the particle center, respectively. This result is in qualitative agreement with some experimental results obtained from grafting reaction studies and from particle morphology studies in polystyrene latexes. This work shows that for a variety of polymerization conditions, the free radical anchoring effect results in significant free radical concentration gradients with radial position in the latex particle and has an impact on the emulsion polymerization mechanisms and kinetics.

This free radical distribution function, determined via a Monte Carlo technique, was then incorporated into the two-phase emulsion polymerization kinetics model developed by Nelson and Sundberg [44, 45] to predict the monomer conversion versus time data available in the literature. Thus, the only difference between the kinetic model characterized by the following major governing equations [47] and that of Nelson and Sundberg is the method used for calculation of the average free radical population in each polymer phase.

$$\begin{aligned} R_p(\text{total}) &= R_{p,c} + R_{p,s} \\ &= k_{p,c}[\mathrm{M}]_c[\mathbf{n}_c/(v_c N_{\mathrm{AV}})] + k_{p,s}[\mathrm{M}]_s[\mathbf{n}_s/(v_s N_{\mathrm{AV}})] \end{aligned} \tag{8.2}$$

$$\mathbf{n} = \mathbf{n}_s + \mathbf{n}_c \tag{8.3}$$

$$\mathbf{n}_s = \mathbf{n} P_i \int_0^{(r_s - r_c)/r} \rho(z)\,dz \Big/ \int_0^{r_s/r} \rho(z)\,dz + \mathbf{n}(1 - P_i)\Phi_s \tag{8.4}$$

$$P_i = (\rho_i/N_p)/[\rho_a/N_p + (k_{\mathrm{tr},m}[\mathrm{M}]_p + k_{\mathrm{tr},t}[\mathrm{T}]_p)\,\mathbf{n}] \tag{8.5}$$

where $R_{p,c}$ and $R_{p,s}$, $k_{p,c}$ and $k_{p,s}$, $[\mathrm{M}]_c$ and $[\mathrm{M}]_s$, and $\mathbf{n}_c$ and $\mathbf{n}_s$ are the polymerization rates, the propagation rate constants, the monomer concentrations, and the average numbers of free radicals in the core-phase and in the shell-phase, respectively. The parameters v_c and v_s and r_c and r_s are the volumes and radii of the core-phase and shell-phase, respectively, r is the radius of the reference particle, **n** is the total average number of free radicals per particle, N_{AV} is the Avogadro's number. P_i represents the probability that a given free radical is terminated by an initiator end-group. ρ_i is the rate of generation of initiator radicals in the continuous aqueous phase, ρ_a is the rate of absorption of free radicals by the latex particles, N_p is the number of particles per unit volume of water, $k_{tr,m}$ and $k_{tr,t}$ are the corresponding chain transfer rate constants for monomer and chain transfer agent, and $[M]_p$ and $[T]_p$ are the concentrations of monomer and chain transfer agent in the particles, respectively.

The Smith and Ewart–Stockmayer–O'Toole treatments [48–50] (see Chapter 4) that are widely used to calculate the average number of free radicals per particle (**n**) are based on the assumption that the various components of the monomer-swollen latex particles (e.g., monomer, polymer, free radicals, chain transfer agent, etc.) are uniformly distributed within the particle volume. A latex particle in emulsion homopolymerization of styrene involves uniform distribution of monomer and polymer within the particle volume except perhaps for a very thin layer near the particle surface. In the case of free radicals, this uniform distribution would only hold in a stochastic sense. However, as illustrated in Eq. (8.1), free radicals are not distributed uniformly in the latex particles when water-soluble initiators are used to initiate the free radical polymerization. The assumption of uniform distribution of free radicals in the latex particles would be valid only if the particles are very small or chain transfer reactions are the dominate mechanism for producing free radicals. If such a nonuniform free radical distribution hypothesis is accepted, the very basis of the Smith and Ewart–Stockmayer–O'Toole methods might be questioned. Despite this potential problem, the Stockmayer–O'Toole solutions for the average number of free radicals per particle have been used for kinetic studies of many emulsion polymerization systems. The theories seem to work reasonably well and have been tested extensively with monomers such as styrene.

Second-stage monomer is expected to be distributed nearly uniformly in the two-phase latex particle volume unless it is not a good solvent for one polymer. One also expects a rather uniform distribution of polymer but a rather complete segregation of the two polymer phases (perfect core-shell segregation is assumed herein). The distribution of free radicals in the two-phase latex particles should be similar to that in the particles of styrene homopolymerization systems. One can argue that the Stockmayer–O'Toole approach would not apply for determination of the total average number of free radicals per particle with such two-phase emulsion polymerization systems. The same arguments, however, would apply to emulsion polymerization of single monomer since the free radicals are not distributed uniformly throughout the latex particle. Thus, one should be able to use the Stockmayer–O'Toole solutions for the two-phase emulsion polymerization systems with the same accuracy as for single-monomer polymerization systems. However, it is not straightforward to apply the Stockmayer–O'Toole solutions to two-phase emulsion polymerization systems. If the bimolecular termination rate constants in each phase are equal, one can employ the Stockmayer–O'Toole solutions directly to calculate the total average number of free radicals per particle because the whole two-phase latex particle is defined as the system when making a mole balance of the free radicals in the polymerization system. On the other hand, some modifications for the Smith and Ewart–Stockmayer–O'Toole treatments are required for two-phase emulsion polymerization systems if these termination rate constants have different magnitudes.

The termination rate constant is dependent on monomer conversion level, polymerization temperature, and the glass transition temperature of the reacting fluid. Liao and Sundberg [51] studied the seeded emulsion polymerization of methyl methacrylate. They concluded that the polymer chain length could also have a significant influence on the termination rate constant (termed the chain entanglement effect). As mentioned above, monomer is distributed nearly uniformly within the two-phase latex particle volume. Thus, for an isothermal reaction, the monomer conversion level and temperature are irrelevant to the question whether the Stockmayer–O'Toole expressions are applicable to two-phase emulsion polymerization systems or not. The two polymer phases were assumed to separate completely into a distinct core-shell particle morphology because of the natural incompatibility of most polymer pairs. These two segregated polymer phases generally possess different glass transition temperatures and polymer molecular weights and, as a consequence, would be expected to have different termination rate constants. In this case, the Stockmayer–O'Toole treatments are not applicable to two-phase emulsion polymerization kinetics. Individual mole balances of the free radicals must be established for the aqueous, core, and shell phases in order to determine the distribution of free radicals among these phases.

The experimental data of the two-phase emulsion polymerization systems of polystyrene/polymethyl methacrylate and polymethyl methacrylate/polystyrene (postformed polymer/preformed polymer) available in the literature [42] were used to test the kinetic model based on the nonuniform distribution of free radicals within the particle volume. This pair of polymers has similar glass transition temperatures and calculated number-average degree of polymerization if the chain transfer reaction is the predominant form of polymer chain length control. Chain transfer reactions (to monomer, chain transfer agent, etc.) play an important role in emulsion polymerization. Such reactions generate small, mobile free radicals. The distribution of such free radicals in polymerizing latex particles is dependent on the location in which these free radicals form and the viscosity of the reaction medium. The free radicals with relatively high mobility should be able to move rapidly in the latex particles. Thus, free radicals generated by chain transfer reactions

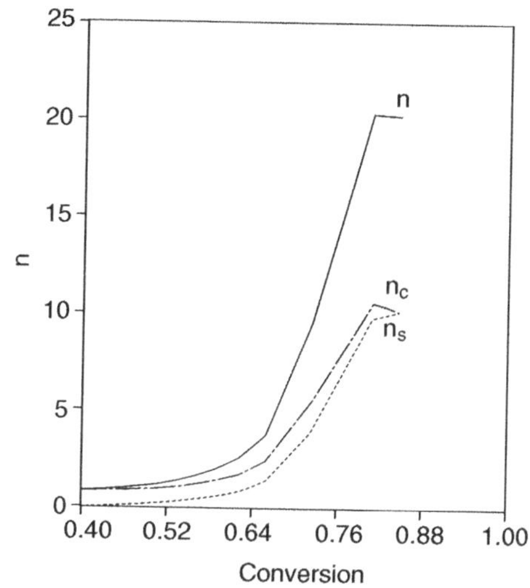

Figure 8.2. Profiles of the calculated average number of free radicals per particle (**n**, $\mathbf{n}_c$ or $\mathbf{n}_s$) as a function of monomer conversion for the experiment at 50 °C with the weight fractions of polystyrene seed particles (154 nm in diameter), the second-stage monomer (methyl methacrylate), and the weight fraction of initiator (potassium persulfate) equal to 0.06, 0.09, and 9.62 $\times 10^{-4}$, respectively.

were assumed to be distributed uniformly within the particle volume. Diffusion-controlled termination and propagation reactions were also incorporated into the model.

The representative computer simulation results are illustrated in Figures 8.2–8.5. Figure 8.2 shows the profiles of the calculated average number of free radicals per particle (**n**, $\mathbf{n}_c$, or $\mathbf{n}_s$) as a function of monomer conversion for the experiment at 50 °C with the weight fractions of polystyrene seed particles (154 nm in diameter), the second-stage monomer (methyl methacrylate), and the weight fraction of initiator (potassium persulfate) equal to 0.06, 0.09, and 9.62×10^{-4}, respectively. The gel effect is significant; **n** increases with increasing monomer conversion throughout most of the seeded emulsion polymerization. After about 80% monomer conversion, **n** starts to level off and even decrease slightly. This is because the reaction system reaches the point corresponding to the critical fractional free volume and the latex particle actually shrinks during polymerization. At the beginning of polymerization, the core-phase predominates in the reaction kinetics because the shell-phase does not exist. At higher monomer conversion, the shell-phase continues to grow and begins to compete effectively with the core-phase for polymerizing methyl methacrylate. This competition does not stop until the free radical polymerization is completed. This is not in agreement with the conclusion of Nelson and Sundberg [45].

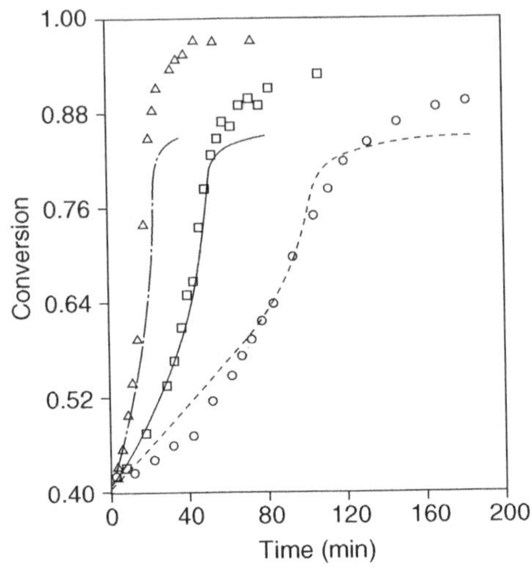

Figure 8.3. Profiles of the monomer conversion (X) as a function of time for the experiments with different initiator concentrations. Weight fraction of initiator: (△) 4.32×10^{-3}, (□) 9.62×10^{-4}, (○) 4.32×10^{-4}. The continuous lines represent the theoretical predictions, and the discrete points represent the experimental data taken from reference 42.

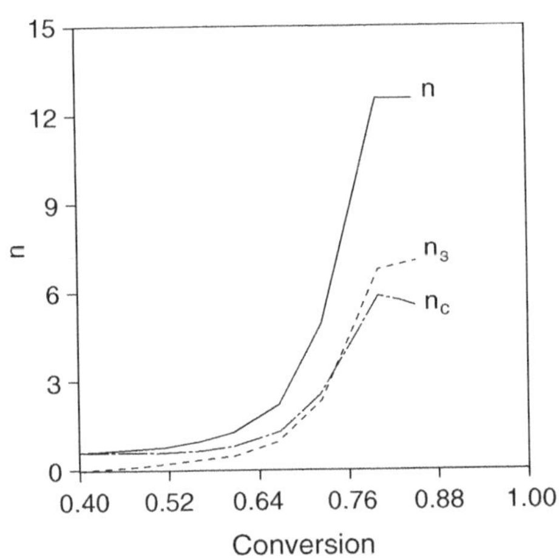

Figure 8.4. Profiles of the calculated average number of free radicals per particle (**n**, $\mathbf{n}_c$, or $\mathbf{n}_s$) as a function of monomer conversion for the experiment at 50 °C with the weight fractions of poly(methyl methacrylate) seed particles (122 nm in diameter), the second-stage monomer (styrene), and the weight fraction of initiator (potassium persulfate) equal to 0.06, 0.09, and 3.72×10^{-4}, respectively.

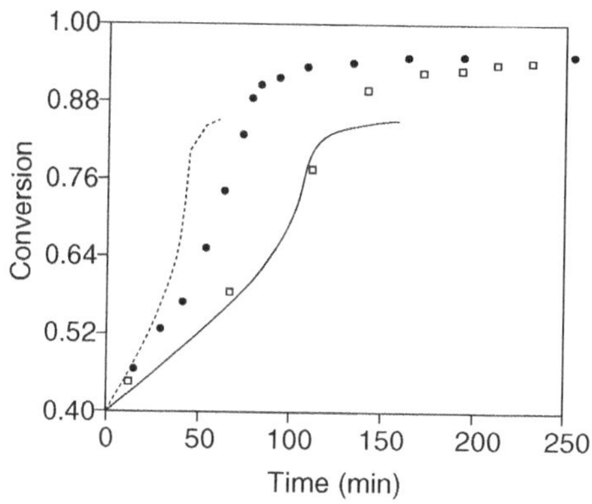

Figure 8.5. Profiles of the monomer conversion (*X*) as a function of time for the experiments with different initiator concentrations. Weight fraction of initiator: (●) 3.43×10^{-3}, (□) 3.72×10^{-4}. The continuous lines represent the theoretical predictions, and the discrete points represent the experimental data taken from reference 42.

Figure 8.3 shows the profiles of the monomer conversion as a function of time for the experiments with different initiator concentrations. Other experimental conditions were kept constant in this series of polymerizations. The only adjustable parameter is the termination rate constant at the very beginning of polymerization and it has a reasonable value of 7×10^5 liter $mol^{-1} s^{-1}$. As expected, the polymerization rate increases with increasing initiator concentration. The kinetic model predicts the polymerization rate data reasonably well. The predicted limiting monomer conversion is lower than the experimental data.

Figure 8.4 shows the profiles of the calculated average number of free radicals per particle as a function of monomer conversion for the experiment at 50 °C with the weight fractions of poly(methyl methacrylate) seed particles (122 nm in diameter), the second-stage monomer (styrene), and the weight fraction of initiator (potassium persulfate) equal to 0.06, 0.09, and 3.72×10^{-4}, respectively. Similar results are observed. Figure 8.5 shows the profiles of the monomer conversion as a function of time for the experiments with different initiator concentrations. Other experimental conditions were kept constant in this series of polymerizations. The only adjustable parameter is the termination rate constant at the very beginning of polymerization and it has a value of 3×10^5 liter $mol^{-1} s^{-1}$. The kinetic model overpredicts the polymerization rate data for the experiment with the weight fraction of initiator equal to 3.43×10^{-3}, but it does adequately predict the trend.

Chern and Poehlein [52] developed a kinetic model based on the nonuniform free radical distribution function to predict the grafting efficiency of the emulsion emulsion polymerization of styrene in the presence of polybutadiene seed latex particles. The predominant grafting reaction appears to be the attack of growing polystyrene chains on the allyl hydrogen atoms of polybutadiene. The results further reinforce the hypothesis that the entering oligomeric radicals are not distributed uniformly within the latex particle volume. De la Cal et al. [53] and Mills et al. [54] performed similar calculations, using somewhat different approaches, and made similar conclusions.

Recently, Durant et al. [55] developed a mechanistic model based on the classic Smith-Ewart theory [48] for the two-phase emulsion polymerization kinetics. This model, which takes into consideration complete kinetic events associated with free radicals, provides a delicate procedure to calculate the polymerization rate for latex particles with two distinct polymer phases. It allows the calculation of the average number of free radicals for each polymer phase and collapses to the correct solutions when applied to single-phase latex particles. Several examples were described for latex particles with core-shell, inverted core-shell, and hemispherical structures, in which the polymer glass transition temperature, monomer concentration and free radical entry rate were varied. This work illustrates the important fact that morphology development and polymerization kinetics are coupled processes and need to be treated simultaneously in order to develop a more realistic model for two-phase emulsion polymerization systems. More efforts are required to advance our knowledge in this research field.

REFERENCES

1. M. Okubo, Y. Katsuta, and T. Matsumoto, *J. Polym. Sci., Polym. Chem. Ed.* **18**, 481 (1980).
2. M. Okubo, A. Yamata, and T. Matsumoto, *J. Polym. Sci., Polym. Chem. Ed.* **18**, 3219 (1980).
3. D. I. Lee, in *Emulsion Polymers and Emulsion Polymerization*, D. R. Bassett and A. E. Hamielec (Eds.), ACS Symposium Series, No. 165, American Chemical Society, Washington, D.C., 1981, p. 405.
4. M. Okubo, M. Seike, and T. Matsumoto, *J. Appl. Polym. Sci.* **28**, 383 (1983).
5. T. I. Min, A. Klein, M. S. El-Aasser, and J. W. Vanderhoff, *J. Polym. Sci., Polym. Chem. Ed.* **21**, 2845 (1983).
6. I. Cho and K. W. Lee, *J. Appl. Polym. Sci.* **30**, 1903 (1985).
7. N. Sutterlin, *Makromol. Chem. Suppl.* **10/11**, 403 (1985).
8. A. T. Skjeltorp, J. Ugelstad, and T. Ellingsen, *J. Colloid Interface Sci.* **113**, 577 (1986).
9. M. P. Merkel, V. L. Dimonie, M. S. El-Aasser, and J. W. Vanderhoff, *J. Polym. Sci., Polym. Chem. Ed.* **25**, 1219 (1987).
10. M. A. Winnik and C. L. Zhao, *Langmuir* **9**, 2053 (1993).
11. J. Brandrup and E. H. Immergut (Eds.), *Polymer Handbook*, 3rd ed., Wiley-Interscience, New York, 1989, p. VII.
12. P. A. Small, *J. Appl. Chem.* **3**, 71 (1953).
13. C. S. Chern and G. W. Poehlein, *J. Polym. Sci., Polym. Chem. Ed.* **25**, 617 (1987).

14. M. P. Merkel, V. L. Dimonie, M. S. El-Aasser, and J. W. Vanderhoff, *J. Polym. Sci., Polym. Chem. Ed.* **25**, 1219 (1987).
15. D. C. Sundberg, A. P. Casassa, J. Pantazopoulos, M. R. Muscato, B. Kronberg, and J. Berg, *J. Appl. Polym. Sci.* **41**, 1425 (1990).
16. Y. C. Chen, V. Dimonie, and M. S. El-Aasser, *Macromolecules* **24**, 3779 (1991).
17. Y. Durant and D. C. Sundberg, *J. Appl. Polym. Sci.* **58**, 1607 (1995).
18. L. J. Gonzalez and J. M. Asua, *Macromolecules* **28**, 3135 (1995).
19. L. J. Gonzalez and J. M. Asua, *Macromolecules* **29**, 383 (1996).
20. L. J. Gonzalez and J. M. Asua, *Macromolecules* **29**, 4520 (1996).
21. A. M. Aerdts, G. Groeninckx, H. F. Zirkzee, H. A. M. van Aert, and J. M. Geurts, *Polymer* **38**, 4247 (1997).
22. J. Stubbs, O. Karlsson, J. E. Jonsson, E. Sundberg, Y. Durant, and D. C. Sundberg, *Colloid Surf. A: Physicochem. Eng. Aspects* **153**, 255 (1999).
23. O. J. Karlsson, H. Hassander, and D. Colombini, *C. R. Chim.* **6**, 1233 (2003).
24. J. M. Stubbs and D. C. Sundberg, *J. Appl. Polym. Sci.* **91**, 1538 (2004).
25. J. M. Stubbs and D. C. Sundberg, *J. Appl. Polym. Sci.* **102**, 945 (2006).
26. J. M. Stubbs and D. C. Sundberg, *J. Appl. Polym. Sci.* **102**, 2043 (2006).
27. C. J. Ferguson, G. T. Russell, and R. G. Gilbert, *Polymer* **44**, 2607 (2003).
28. K. Zhao, P. Sun, D. Liu, and G. Dai, *Eur. Polym. J.* **40**, 89 (2004).
29. M. Pan, L. Zhang, L. Wan, and R. Guo, *Polymer* **44**, 7121 (2003).
30. Y. G. Durant and D. C. Sundberg, *Polym. React. Eng.* **11**, 433 (2003).
31. K. Landfester and H. W. Spiess, *Acta Polymerica* **49**, 451 (1998).
32. S. Kirsch, K. Landfester, O. Schaffer, and M. S. El-Aasser, *Acta Polymerica* **50**, 347 (1999).
33. J. G. Tsavalas, F. J. Schork, and K. Landfester, *JCT Res.* **1**, 53 (2004).
34. D. J. Hourston, H. X. Zhang, M. Song, H. M. Pollock, and A. Hammiche, *Thermochim. Acta* **294**, 23 (1997).
35. D. J. Hourston, M. Song, and Y. Pan, *J. Brazilian Chem. Soc.* **12**, 87 (2001).
36. J. M. Stubbs and D. C. Sundberg, *Polymer* **46**, 1125 (2005).
37. J. M. Stubbs and D. C. Sundberg, *J. Polym. Sci., Polym. Phys. Ed.* **46**, 1125 (2005).
38. T. Araki, H. Ade, J. M. Stubbs, D. C. Sundberg, G. E. Mitchell, J. B. Kortright, and A. L. D. Kilcoyne, *Appl. Phys. Lett.* **89**, 124106 (2006).
39. D. C. Sundberg and Y. G. Durant, *Polym. React. Eng.* **11**, 379 (2003).
40. S. L. Rosen, *J. Appl. Polym. Sci.* **17**, 1805 (1973).
41. C. W. Chiu, *The Kinetics of Two-Phase Emulsion Polymerization*, PhD Thesis, Carnegie-Mellon University, Pittsburgh, PA, 1978.
42. P. Varshney, *Two Phase Emulsion Polymerization Kinetics of Styrene and Methyl-Methacrylate*, MS Thesis, Department Of Chemical Engineering, University Of New Hampshire, Durham, NH, 1981.
43. W. V. Smith and R. H. Ewart, *J. Chem. Phys.* **16**, 592 (1948).
44. D. Nelson, *Kinetics of Two Phase Emulsion Polymerization*, MS Thesis, Department of Chemical Engineering, University of New Hampshire, Durham, NH, 1983.

45. D. Nelson and D. C. Sundberg, *Kinetics of Two Phase Emulsion Polymerization*, In AIChE Manuscript No. 7739, AIChE Meeting, Houston, TX, March, 1983.
46. A. R. H. Cornish, *Trans. Inst. Chem. Eng.* **43**, T333 (1965).
47. C. S. Chern and G. W. Poehlein, *J. Polym. Sci., Polym. Chem. Ed.* **28**, 3055 (1990).
48. W. V. Smith and R. H. Ewart, *J. Chem. Phys.* **16**, 592 (1948).
49. W. H. Stockmayer, *J. Polym. Sci.* **24**, 314 (1957).
50. J. T. O'Toole, *J. Appl. Polym. Sci.* **9**, 1291 (1965).
51. C. T. Liao and D. C. Sundberg, *Polymer* **27**, 265 (1986).
52. C. S. Chern and G. W. Poehlein, *J. Polym. Sci., Polym. Chem. Ed.* **28**, 3073 (1990).
53. J. C. de la Cal, R. Urzay, A. Zamora, J. Forcada, and J. M. Asua, *J. Polym. Sci., Polym. Chem. Ed.* **28**, 1011 (1990).
54. M. F. Mills, R. G. Gilbert, and D. H. Napper, *Macromolecules* **23**, 4247 (1990).
55. Y. G. Durant, R. Carrier, and D. C. Sundberg, *Polym. React. Eng.* **11**, 433 (2003).

9

APPLICATIONS OF EMULSION POLYMERS

Conventional solvent-borne polymer systems continue to lose ground under the increasing pressure of environmental protection. Water-borne emulsion polymers such as polystyrene, polyvinyl chloride, polyvinyl acetate, acrylic copolymers, styrene–acrylic copolymers, vinyl acetate–acrylic copolymers, vinyl acetate–vinyl chloride copolymers, ethylene–vinyl acetate copolymers, styrene–butadiene copolymers, acrylonitrile–butadiene–styrene copolymers, and so on, can be used to replace the polymer systems containing organic solvents (total solids content ≤60%). These latex products find a variety of industrial applications ranging from synthetic rubbers, adhesives, binders, trade paints, industrial coatings, printing inks, thermoplastics, and emulsion aggregation toners to immunoassay products based on the affinity interaction between latex particles and biomolecules. Styrene–butadiene copolymers and polyvinyl acetate for emulsion paints, for example, were introduced around 1946–1950. The versatile acrylic polymers comprising monomeric units with a wide range of glass transition temperature and polarity can display excellent mechanical properties, adhesion properties, optical properties, and ultraviolet light or hydrolysis resistance. These acrylic latex products were first launched in the emulsion paint marketplace in 1951. Other alternative technologies that compete with traditional volatile organic compound (VOC) products include water-based polyurethane dispersions, solvent-based high-solids-content polymer systems, UV-curable coatings, and powder coatings.

Despite the favorable atmosphere for the environmentally friendly polymer systems, formulation chemists must learn to work with latex products and very often they will find it very difficult to handle the water-based coating and

Principles and Applications of Emulsion Polymerization, by Chorng-Shyan Chern

printing systems. End-users also need to accept the fact that these latex products have limitations, and perhaps they can never achieve the excellent performance properties offered by the solvent-based counterparts, not to mention the problems often associated with water such as the instability of latex products during manufacturing, storage, or transportation (e.g., chemical, mechanical, freeze–thaw, and heat stability), bacterial contamination, sedimentation or creaming, foaming, poor wetting, slow drying, and unsatisfactory film formation.

The colloidal stability issue associated with the extremely large oil–water interfacial area of latex particles has been covered in Chapter 2. Other important factors that affect the ultimate performance properties of water-based coating and printing systems are covered in this chapter, even though some of these critical issues might not be directly related to the emulsion polymers themselves.

9.1 PHYSICAL PROPERTIES OF EMULSION POLYMERS

The mechanical properties (e.g., hardness, flexibility, impact resistance, abrasion resistance, and scratch resistance) of polymers are primarily characterized by the polymer molecular weight and molecular weight distribution, polymer morphology (e.g., semicrystalline and amorphous structures), and crosslinking reactions. The purpose of this section is to give the reader an introductory background about the important factors that affect the physical properties of emulsion polymers.

9.1.1 Effect of Polymer Molecular Weight

Polymer molecular weights and molecular weight distributions obtained from conventional emulsion polymerization are covered in Chapter 4, Section 4.5. In the absence of chain transfer agent, significant transfer to monomer, or crosslinking agent, increasing the number of latex particles per unit volume of water (i.e., increasing the degree of segregation of free radicals among the discrete reaction loci) and decreasing the concentration of initiator result in an increase in the molecular weight of emulsion polymer. However, incorporation of chain transfer agent or crosslinking agent into the emulsion polymerization system is perhaps the most effective tool to control the evolution of polymer molecular weight and molecular weight distribution.

Polymers with very low molecular weight are viscous liquids if their glass transition temperatures are below the ambient temperature. At higher molecular weight, these polymeric materials become cheesy elastomers with low tensile strength and elongation to break. At still higher molecular weight (on the order of $10^5\,\mathrm{g\,mol^{-1}}$ and higher), the polymer chains become entangled enough to show true rubbery behavior to short-term deformations, and the elongation to break is found to be on the order of 10^3 percent. The chain

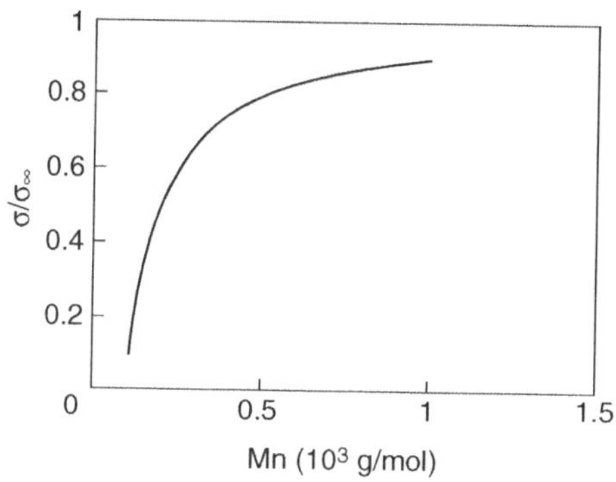

Figure 9.1. Calculated dimensionless tensile strength (σ/σ_∞) of polymers as a function of number-average molecular weight (M_n) according to Eq. (9.1). An arbitrary value of 100 g mol^{-1} is assigned to K/σ_∞.

entanglement effect is significant only when the polymer chain length reaches a critical value. On the other hand, polymers with even very low molecular weight are quite brittle when their glass transition temperatures are well above ambient temperature [1].

Above the minimum polymer molecular weight required to form a testing specimen, both the tensile strength and elogation of polymers with very high molecular weight increase toward a limiting value [2–5]. Polar polymers and polymers with hydrogen bonding between polymer chains reach their maximum properties at lower molecular weight than do nonpolar polymers. In most cases, the tensile strength of polymers follows an equation of the following form [6]:

$$\sigma = \sigma_\infty - K/M_n \quad \text{or} \quad \sigma/\sigma_\infty = 1 - (K/\sigma_\infty)/M_n \tag{9.1}$$

where σ is the tensile strength of polymers, σ_∞ is the limiting tensile strength at infinite polymer chain length, M_n is the number-average polymer molecular weight, and K is a constant. As a rule of thumb, the physical properties of polymers first increase rapidly and then level off as the polymer molecular weight is increased, as demonstrated in Figure 9.1.

9.1.2 Effect of Polymer Morphology

Polymers can be considered simply as large molecules comprising repeating structural units. The arrangement of repeating structural units can have a significant influence on the physical properties of a polymeric material. Most polymers are amorphous or have an amorphous-like component even if they are crystalline. The latter type are termed semicrystalline polymers. Crystalline microdomains are those with a regular, ordered packing of molecular units.

Polymers that can be packed into crystalline microstructures have essentially linear chains in which substituents or side-chain groups are small enough to fit into an orderly arrangement (e.g., polyethylene, polyvinyl alcohol, polyvinyl fluoride, etc.) or are disposed regularly and symmetrically along the chain (e.g., polyesters, polyamides, etc.). In contrast, amorphous polymers are those with bulky side groups and irregular configurations that prohibit these macromolecules from being packed into crystalline microstructures [e.g., polyvinyl acetate, polystyrene, poly(methyl methacrylate), etc.]. Semicrystalline polymers are those with rigid crystalline regions interconnected by flexible amorphous or disordered regions. The crystalline and amorphous microdomains are characterized by melting point (a first-order transition from the crystalline phase to the amorphous phase) and glass transition (a second-order transition from the glassy state to the rubbery state or vice versa), respectively. In general, semicrystalline polymers with high crystallinity exhibit excellent physical and chemical properties as compared to amorphous polymers. However, highly crystalline polymers are seldom used in coatings because of their very poor solubility in solvents. Because most emulsion polymers are amorphous in nature, we will focus our attention on these types of materials hereinafter.

The glass transition temperature (T_g) is defined as a region in which a transition from the hard, brittle glassy state to the soft, flexible rubbery state occurs when the temperature is continuously increased. At the glass transition temperature, polymers show abrupt changes in many physical properties. For example, the elastic modulus may decrease by a factor of over 1000 as the ambient temperature is raised through the glass transition region. Thus, the glass transition temperature is perhaps the most important characteristic of emulsion polymers, as far as mechanical properties are concerned. Some changes in the polymer properties such as density, specific heat, and refractive index are relatively insensitive to the rate at which the phase transition occurs. On the other hand, other characteristics such as flexibility, elastic modulus, rheological properties, and dielectric properties are dependent on the phase transition rate. Figure 9.2 shows a schematic representation of changes in specific volume with temperature for amorphous or semicrystalline polymers [7]. Figure 9.3 illustrates a schematic representation of the dependence of tensile modulus of polymers on temperature [8]. Changes in the type of mechanical behavior of polymers with temperature are also included in this generalized plot. The glass transition temperature is not a true constant; it is strongly dependent on the time scale of the measurements. The glass transition temperature is generally determined by experiments that correspond to a time scale of seconds or minutes. Table 9.1 lists the glass transition temperature data for common homopolymers involved in emulsion polymerizations.

Similar to many mechanical properties [e.g., tensile strength, see Eq. (9.1) and Figure 9.1], the effect of polymer molecular weight on the glass transition temperature can be expressed by the following equation:

$$T_g = T_{g\infty} - K/M_n \tag{9.2}$$

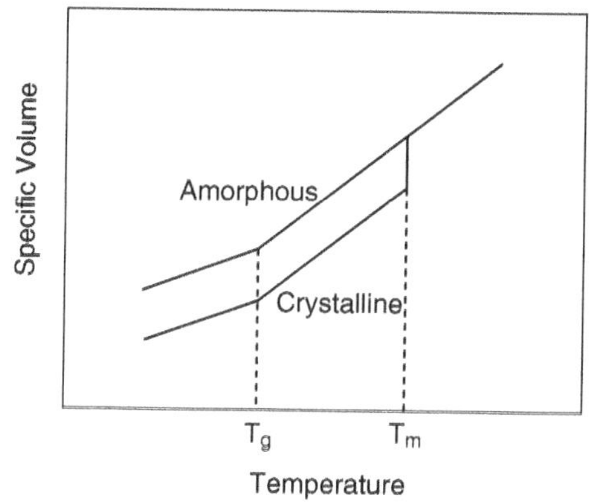

Figure 9.2. A schematic representation of changes in specific volume with temperature for amorphous or semicrystalline polymers.

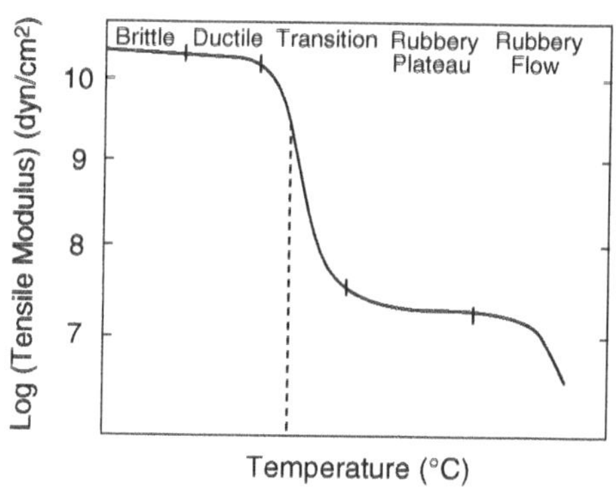

Figure 9.3. A schematic representation of the dependence of tensile modulus of polymers on temperature [8].

where $T_{g\infty}$ is the asymptotic value of the glass transition temperature at infinite polymer chain length. The glass transition temperature of polymers first increases rapidly and then levels off with increasing polymer molecular weight. However, manipulation of polymer molecular weight alone is apparently ineffective in designing latex products of vast dimensions.

Emulsion copolymerization of monomers with different glass transition temperatures (Table 9.1) is the technique of choice to control the mechanical properties of polymers. For random copolymers, the glass transition temperature (T_g) can be estimated with the following equation:

$$1/T_g = \sum_i w_i/T_{gi} \tag{9.3}$$

where T_{gi} and w_i are the glass transition temperature and weight fraction of component i in the copolymer, respectively, and $\Sigma w_i = 1$.

Table 9.1. Glass Transition Temperature Data for Common Emulsion Polymers

Homopolymers	T_g (°C)
Poly(acrylic acid)	106
Poly(methyl acrylate)	8
Poly(ethyl acrylate)	–22
Poly(isopropyl acrylate)	–5
Poly(*n*-propyl acrylate)	–52
Poly(isobutyl acrylate)	–40
Poly(*n*-butyl acrylate)	–54
Poly(*t*-butyl acrylate)	43
Poly(2-ethyl acrylate)	–85
Poly(*n*-octyl acrylate)	–80
Poly(methacrylic acid)	228
Poly(methyl methacrylate)	105
Poly(ethyl methacrylate)	65
Poly(isopropyl methacrylate)	81
Poly(*n*-propyl methacrylate)	33
Poly(isobutyl methacrylate)	48
Poly(*n*-butyl methacrylate)	20
Poly(*t*-butyl methacrylate)	107
Poly(*n*-octyl methacrylate)	–20
Poly(isobornyl methacrylate)	–114
Poly(ethylene glycol dimethacrylate)	132
Poly(2-hydroxyethyl methacrylate)	55
Poly(glycidyl methacrylate)	60
Polyethylene (high molecular weight)	–130
Polybutadiene (random)	–85
Polybutadiene (high *cis*)	–102
Poly(vinyl acetate)	30
Poly(vinyl chloride)	82
Polyvinyl pyrrolidone	54
Polystyrene	100
Poly(α-methyl styrene)	180
Polyacrylonitrile	125

These empirical relationships along with the glass transition temperature data established for homopolymers (Table 9.1) allow polymer chemists to determine the optimal position of the glass transition temperature and then design adequate emulsion polymer compositions to fulfill end-users' requirements. *n*-Butyl acrylate (T_g = –54 °C) and 2-ethylhexyl acrylate (T_g = –85 °C), for example, are widely used as the major components of water-based pressure-sensitive adhesives. Vinyl acetate and *n*-butyl acrylate copolymer latexes with a weight ratio of about 80:20 (T_g = 8 °C) are a primary choice for interior architectural coatings, where resistance to ultraviolet light is not an issue.

In addition to conventional surfactants, hydroxyethyl cellulose (HEC) is generally used as a protective colloid that imparts excellent colloidal stability and rheological properties to the latex products. In contrast, *n*-butyl acrylate and methyl methacrylate copolymers with a weight ratio of approximately 48:52 ($T_g = 7\,°C$) are very suitable for the exterior latex paint applications. A small amount of acrylic acid or methacrylic acid (<5% based on total monomer weight) is generally incorporated into the latex particles to enhance the colloidal stability and adhesion properties. Polyvinyl acetate ($T_g = 30\,°C$) latexes using polyvinyl alcohol as the protective colloid are useful for laminating adhesive applications. One major process used to manufacture polyvinyl chloride ($T_g = 82\,°C$) powders is semibatch emulsion polymerization. Recently, more emphasis has been placed on controlling the shape and especially the breadth of the glass transition region of polymeric materials. This versatile, delicate approach extends the performance properties of emulsion polymers to a much higher level.

9.1.3 Effect of Crosslinking Reactions

Crosslinking reactions arising from the incorporation of a small amount of multifunctional monomers such as divinyl benzene and ethylene glycol dimethacrylate into emulsion polymerization systems result in polymers showing a quite different characteristic property curve compared to that illustrated in Figure 9.3 [9, 10]. The crosslink points within the polymeric matrix, even at relatively low levels, eliminate the rubbery flow region because these junction points effectively prevent the long range, independent translational movements of polymer chains required for flow. The rubber plateau is shifted upwards when the crosslinking density increases. This is because the average length of chain segments between two crosslink points in the polymeric matrix is reduced. As a result, stronger force is required to elongate the testing sample. However, increased crosslinking density has a much smaller effect on tensile modulus of polymers in the glassy state than in the rubbery plateau region because the shorter range cohesive forces that oppose the elongation of glassy materials are not dependent on the average length of chain segments between two crosslink-points in the polymeric matrix.

Comparing an uncrosslinked emulsion polymer film and a highly crosslinked automotive top coat at temperatures well below their respective glass transition temperatures as an example, both polymeric films have tensile modulus values of approximately $2 \times 10^{10}\,\text{dyn}\,\text{cm}^{-2}$ [8]. Nevertheless, these two polymer systems show very different mechanical properties in the rubbery plateau region. The tensile modulus values are approximately $(5\text{–}10) \times 10^6$ and $(2\text{–}8) \times 10^8\,\text{dyn}\,\text{cm}^{-2}$ for the emulsion polymer and automotive top coat, respectively. Thus, one can certainly improve the physical properties of this latex product at temperatures higher than its glass transition temperature by incorporating a small quantity of crosslinking agents into the polymerization system. However, the resultant crosslink points within the polymer particles may hinder the coalescence of the approaching particles and the subsequent mixing of polymer chains belonging to different particles. In this case, a continuous, tough polymeric film cannot be achieved during the film formation process. Thus, the ultimate performance properties of the crosslinked emulsion polymer simply cannot be realized. This scenario illustrates the fact that film formation of latex particles plays an important role in determining the final performance properties of the resultant polymeric film. This subject will be further discussed later (see Section 9.3).

9.2 RHEOLOGICAL PROPERTIES OF EMULSION POLYMERS

Rheological properties of latex products reflect the deformation or flow of emulsion polymers subjected to shear forces [11, 12]. Rheology has significant effects on mixing and heat transfer during emulsion polymerization, as well as on postpolymerization processing such as the transport of latex products, storage, handling, and mixing of emulsion polymers with cosolvents and additives (e.g., defoamers, wetting agents, leveling agents, wax emulsions, rheological modifiers, bactericides, etc.) to manufacture water-based coating and printing products. Rheology is also important in the application of the finished coatings products to a variety of substrates (e.g., brush coating, roll coating, and spray coating) and the subsequent film formation process. For instance, emulsion polymerization is an extremely exothermic process; therefore, heat transfer is a critical issue that needs to be taken into consideration in the design and operation of reactors. The very high viscosity of the semibatch emulsion polymerization system significantly reduces the rate of heat transfer. Under this circumstance, the risk of having a thermal runaway situation (or even failure of the batch) can be greatly increased. Adding defoamers with very low HLB values to water-borne coatings, for example, normally requires sufficient mixing to ensure the quite hydrophobic molecules be uniformly distributed within the continuous aqueous phase. Otherwise, these defoamers may not only function improperly but also produce defects in the resultant surface coating films.

In general, latex products with values of Deborah number close to unity exhibit viscoelastic behavior, which strongly depends on the hydrodynamic interactions between the polymer particles and water and the interparticle interactions. The viscosity of a latex product is a rheological property that measures the resistance to flow in response to the applied shear force. It increases exponentially with increasing total solids content of the emulsion polymer. At constant total solids content, the viscosity of the colloidal dispersion increases significantly with decreasing particle size. Furthermore, latex products generally show a shear-thinning behavior; viscosity decreases with

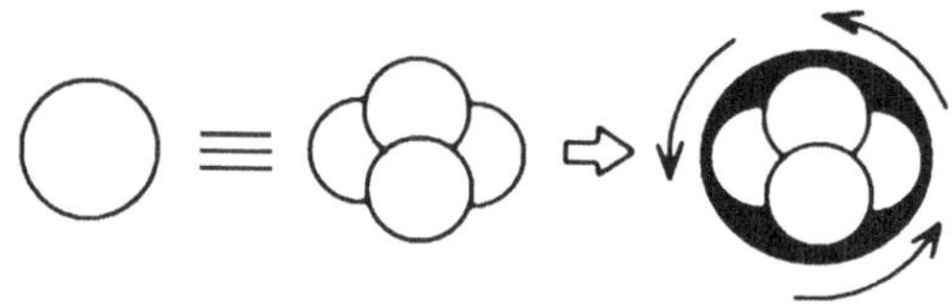

Figure 9.4. A schematic representation of the solid polymer volume versus swept hydrodynamic volume for a tetralobed latex particle [13].

increasing shear rate. For common coatings applications, adequately designed rheological properties of latex products are required for the formation of polymeric films with excellent appearance and physical and chemical properties.

The versatile Rhoplex AC-388 (a copolymer of methyl methacrylate and *n*-butyl acrylate, Rohm and Haas) has been a very successful acrylic latex paint binder for the market of exterior trade paints. It was designed primarily for formulations of durable outside house and trim paints for wood and masonry. The unique rheological properties of Rhoplex AC-388 provide paint formulation chemists with superior leveling and film build. This latex product also offers excellent color acceptance and tint retention, good wet adhesion and chalk adhesion, and good dirt resistance. In the 1980s Rohm and Haas, based on a novel technology for preparing nonuniform latex particles, started to promote an even more interesting acrylic latex paint binder, Rhoplex Multilobe 100 [13, 14]. Figure 9.4 shows the difference between the actual polymer volume and the effective hydrodynamic volume provided by a latex particle comprising four lobes in a tetrahedral configuration [13]. For comparison, a spherical latex particle with equal polymer volume is also included in this schematic diagram. According to Chou et al. [13], the tetrahedral lobed polymer particles in water can sweep out a hydrodynamic volume about 2.5 times that of spherical particles having exactly the same solid volume as the lobed particles. A lobed polymer particle whose solid volume is equal to that of a spherical particle with a diameter of 750 nm displays a hydrodynamic volume approximately the same as that of a spherical particle with a diameter of 1000 nm. As a latex paint binder, Rhoplex Multilobe 100 can offer enhanced brushing viscosity without sacrificing the pigment binding capacity.

Some latex products exhibit a unique non-Newtonian flow behavior (e.g., a series of rheology-controlled emulsion polymers, Joncryl 89, 77, and 74, originally developed at S. C. Johnson Wax). The viscosity of these types of latex products first decreases rapidly and then levels off (Newtonian behavior) as the shear rate is increased. These emulsion polymers offer excellent rheological properties, resolubility (for cleaning coating machines), gloss, and hardness, and they are widely used in the market of overprint varnishes.

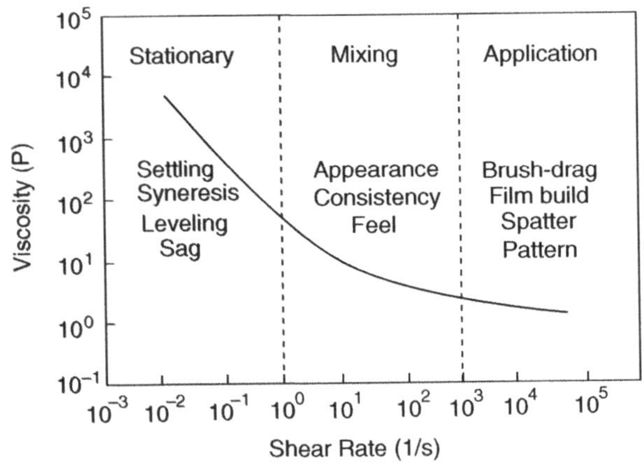

Figure 9.5. A schematic representation of the major role of rheological properties in a typical coating process and some performance properties that are affected by the shear-thinning viscosity profile.

As mentioned above, increasing shear rate generally results in a reduction in the viscosity of a latex product. This shear-thinning behavior can be effectively modified by natural or synthetic rheology modifiers (or thickeners) to meet various application requirements of end-users. Figure 9.5 illustrates the major role of rheological properties in a typical coating process and some performance properties that are affected by the shear-thinning viscosity profile. The most widely used rheology modifiers include cellulosic thickeners, alkali-soluble or -swellable thickeners, and associative thickeners [15]. Latex products containing cellulosic derivatives show strong shear-thinning behavior, which makes it more difficult to produce coatings with balanced film build-up (or hiding) and leveling properties. Furthermore, the chain entanglement of high-molecular-weight polymers in the continuous aqueous phase leads to viscoelastic effects, which cause roller spatter [16–18]. The Polyphobes technology originally developed at Desoto involves hydrophobically modified alkali-swellable polymers, with a molecular weight of about several hundred thousand, that swell significantly in aqueous media upon neutralization [19–22]. Nonionic associative thickeners such as hydrophobically modified ethoxylated urethane polymers most commonly comprise a linear polyethylene oxide backbone and hydrophobic chains distributed along the backbone and at the terminal points of the backbone [23, 24]. Associative rheology modifiers are defined as thickeners that build a physically crosslinked network structure by interacting with themselves and with other formulation components (e.g., polymer particles, pigments, fillers, surfactants, cosolvents, etc.) within a latex paint.

Incorporation of cosolvents into water-based coating and printing systems reduces the efficiency of associative rheology modifiers because of the greatly

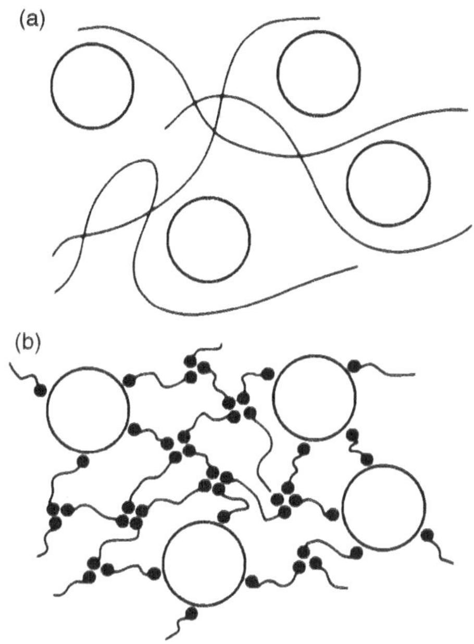

Figure 9.6. A schematic representation of the thickening mechanism of **(a)** conventional water-soluble polymers and **(b)** associative rheology modifiers in a simplified latex paint system. The symbols ○, ——, and •—• represent latex particles, conventional water-soluble polymers, and associative rheology modifier molecules, respectively.

decreased hydrophobic interactions of rheology modifier molecules with themselves and with colloidal particles (polymer particles, pigments, or fillers). This type of rheological modifiers is considered the best in formulating water-based coating and printing materials. The thickening mechanism of conventional water-soluble polymers and associative rheology modifiers in a simplified latex paint system is schematically shown in Figure 9.6.

9.3 FILM FORMATION OF EMULSION POLYMERS

Film formation is a process that converts the wet latex film into a solid polymeric film by evaporation of water after application of the coating material to the object to be coated. A continuous, tough polymeric film can form as a result of the film formation process. On the other hand, a powdery polymeric layer that eventually exhibits no mechanical strength is obtained if the film formation process is incomplete. Satisfactory film formation is a prerequisite for the coating materials to fully exhibit their excellent mechanical properties. Thus, to gain an insight into the mechanism of film formation is essential for assuring the primary performance of latex products.

A latex product is a fine dispersion of an extremely large number of polymer particles in water by using anionic surfactants (electrostatic repulsion mechanism) and/or nonionic surfactants (steric repulsion mechanism) to achieve satisfactory colloidal stability. In order to obtain an integral coating film with desired mechanical properties, these stabilizing actions must be overcome and the individual polymer particles must coalesce into a continuous polymeric film. The latex particles come closer when water is driven off the wet latex film. As these polymer particles approach one another, they can be thought of as forming the walls of capillary tubes. In a capillary tube, the surface tension can result in a force striving to collapse the tube. The force originating from the surface tension increases with a decrease in the diameter of the capillary tube. Coalescence of the latex particles is possible only when the particles are close enough so that the repulsive force is overcome by the capillary tube-collapsing force. More importantly, the polymer molecules within the latex particles must possess the capability of diffusing easily from particle to particle. In this manner, the boundary between two coalescing polymer particles can disappear completely. Thus, the rate of coalescence of latex particles is strongly dependent on the particle size and $(T - T_g)$, where T is the temperature at which the film formation process proceeds. In general, the smaller the latex particle size and the lower the glass transition temperature of the emulsion polymer, the faster the rate of film formation. For more detail information regarding the mechanism of film formation, refer to References 25–31.

A relatively old patent [32] discloses an interesting technology that can be used to produce an internally plasticized polymer latex product. Basically, the latex product can be prepared by a multistage emulsion polymerization process. The first-stage emulsion polymer containing a large amount of functional monomers is highly water-swellable or water-soluble. The second-stage emulsion polymer has a higher glass transition temperature, and it is less hydrophilic than the first-stage polymer. The resultant latex products are expected to have an inverted core-shell particle morphology, and they are useful in coatings, adhesives, and binders.

Devon et al. [33] prepared a series of acrylic latexes with core-shell particle morphologies. The minimum film formation temperatures (MFFT) of the latexes are expected to change with the core-shell characteristics in the following order:

Soft–hard (core-shell) > medium–medium > hard–soft

Interesting enough, the above trend was confirmed experimentally only when the shell thickness was greater than a certain value. Thus, the latex particles with a thinner, softer shell surrounding a harder core required a higher temperature for successful film formation than that with a thicker shell with exactly the same polymer composition because more deformation was required to produce a continuous film for the former case.

Omi et al. [34] investigated the effects of latex products with a crosslinked soft core and a hard shell on the polymer particle morphology and minimum film formation temperature. The latex products were prepared by either a batch emulsion polymerization process (flooded with the second-stage monomer) or a semibatch polymerization process (starved monomer feed). The experimental results show that the minimum film formation temperature data of emulsion polymers obtained from the batch process are greater than those produced by the semibatch process when the crosslinking density is increased. A reasonable correlation between the minimum film formation temperatures of latex products and the molecular weight of the THF-soluble fraction of emulsion polymers can be observed regardless of the mode of polymerization operation. All the examples cited here illustrate an important concept: It is absolutely possible to design an emulsion polymer with a glass transition temperature higher than the corresponding minimum film formation temperature.

When the application temperature is below the minimum film formation temperature, emulsion polymer is incapable of forming an integral film and, as a result, the ultimate performance properties cannot be achieved. Moreover, the undesirable side effects such as crack and greatly reduced gloss, adhesion, water resistance, and durability may appear in the coating film. Making the latex particles softer avoids this problem, but the mechanical properties and stain resistance may be adversely affected. To resolve this dilemma, a small amount (a few weight percent) of organic solvents with relatively high boiling point (termed coalescing agents) can be used in the coating formulations. Coalescing agents are partitioned between the continuous aqueous phase and the polymer particle phase. The improved film formation process is primarily attributed to the plasticization effect provided by the coalescing agents residing inside the latex particles, which reduces the minimum film formation temperature of the polymeric coatings. The extent of plasticization of the polymer particles is controlled by the solubility parameters of coalescing agents and emulsion polymer (see Chapter 2, Section 2.2.3). The smaller the difference between the solubility parameters of coalescing agents and emulsion polymer, the larger the fraction of coalescing agents that can be ultimately incorporated into the polymer particle phase. In addition, the fraction of coalescing agents present in the aqueous phase tends to lower the rate of evaporation of water and, thus, improve the flow and leveling properties of the coating system. All these factors contribute to the formation of polymeric films of excellent quality. However, coalescing solvents are evaporated slowly during and after the film formation process. Under this circumstance, addition of coalescing agents significantly reduces the drying speed, which is undesirable in commercial production. It is therefore necessary for polymer chemists to delicately balance the intrinsic performance properties of emulsion polymers and the efficiency of the coating processes. Coalescing solvents also increase the VOC content of the water-based coating.

Representative coalescing agents include 2,2,4-trimethyl-1,3-pentanediol monoisobutyrate (Texanol), hexanediol, ethyleneglycol monoethyl ether, ethyleneglycol monobutyl ether (butyl cellosolve), diethyleneglycol butyl ether (butyl carbitol), and xylene. As a rule of thumb, a mixture of coalescing agents with different water solubilities and boiling points are recommended for an optimum coating formula.

9.4 FOAMING AND ANTIFOAMING ANGENTS

Foam generation could be extremely troublesome in the manufacture, transport, handling, and application of water-borne coating and printing products. This is due to the increasing demands for ever-faster production rates and application speeds. This inherent problem with water-based coating and printing systems is caused by the very high surface tension (i.e., the water–air interfacial tension) and the presence of impurities (e.g., dodecyl alcohol originating from surfactants such as sodium dodecyl sulfate) that help stabilize the foam. Foams can be considered as a type of emulsion, in which the dispersed phase is a gas. The morphological structure of foams is generally polyhedral; foams consist of bubbles that are nearly polyhedral in shape and have narrow lamellar films of very low curvature separating the dispersed gas phase (Figure 9.7 [35]). In the foam formation process, the incipient dispersed gas phase is present as a bulk or condensed phase. Small volumes of the future dispersed gas phase are introduced into the colloidal system by mechanical agitation during the production of coatings, by pumping during package filling or by shear or spraying during application. When foam is generated, water molecules drain due to the gravitational force, and the closed-pack arrays of anionic end-groups of surfactant approach each other. At some point, the forces come into balance with the magnitude of the electrostatic repulsion force controlling the ultimate thickness of the foam films. The high elasticity, resilience, and surface viscosity of the liquid foam films contribute to the stabilization of foams [36].

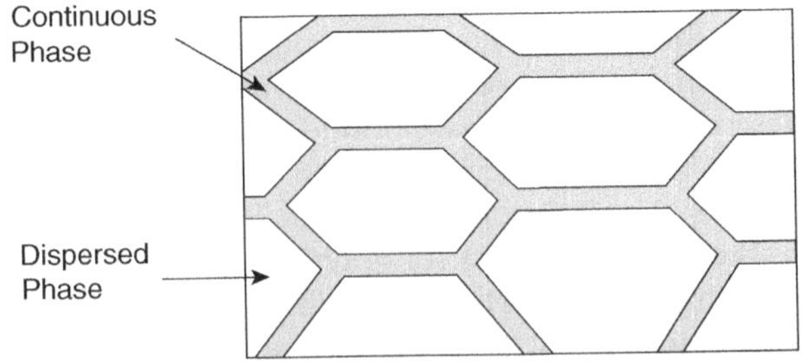

Figure 9.7. Foams consist of closely packed bubbles such that their shape becomes distorted from the spherical one [35].

The collapse of foam is attributed to (a) the diffusion of gas molecules from a small bubble with higher internal pressure to a large one with lower internal pressure or into the bulk gas phase surrounding the foam system, (b) coalescence of bubbles due to capillary flow that results in rupture of the lamellar film between the adjacent bubbles (usually slower than (a) and occurring even in stabilized foam system), and (c) rapid hydrodynamic drainage of liquid between bubbles that leads to rapid collapse of bubbles [35]. In most nonrigid foam systems, all three mechanisms are operative simultaneously to some extent during the foam collapse process.

Use of a very small amount of effective antifoaming agents is beneficial in inhibiting or alleviating a variety of common coating problems such as:

(a) The increase of viscosity and decrease of mechanical shearing power during the milling process, thereby leading to smaller batch sizes and poor quality of the pigment/emulsion polymer dispersion
(b) The increase of volume during the letdown and mixing steps, leading to overflow
(c) Slower package-filling rates due to inefficient pumping
(d) Incorporation of air into coating systems during transport and handling
(e) Slower printing-press speeds or lower pressures during spraying
(f) Some surface defects on polymer-coated substrates, leading to poor appearance, a reduction in gloss, and less protection of substrates

Antifoaming agents (or defoamers) are relatively hydrophobic surface-active species, and they are widely used in water-based coating and printing systems. The action of antifoaming agents in preventing foaming first involves the displacement of the foam-producing surfactant from the gas-solid interface. This is followed by the incorporation of the poor foam-stabilizing characteristics of defoamers into the foam system [35–38]. The defoaming mechanisms include the following [35]:

(a) The antifoaming agent molecules may displace stabilizing surfactant molecules, thereby leading to the break down of bubbles (Figure 9.8a).
(b) The antifoaming agent molecules may displace stabilizing surfactant molecules by spreading as a lens at the gas–water interface (Figure 9.8b).

Reducing the droplet size of the antifoaming composition to approximately the thickness of a foam lamella is a perquisite for its effectiveness as an antifoaming agent. Nevertheless, common defoamers in such a state are unstable and tend to coalesce upon standing. To resolve this problem, silicone-based defoamers are often provided by the suppliers in the form of oil-in-water

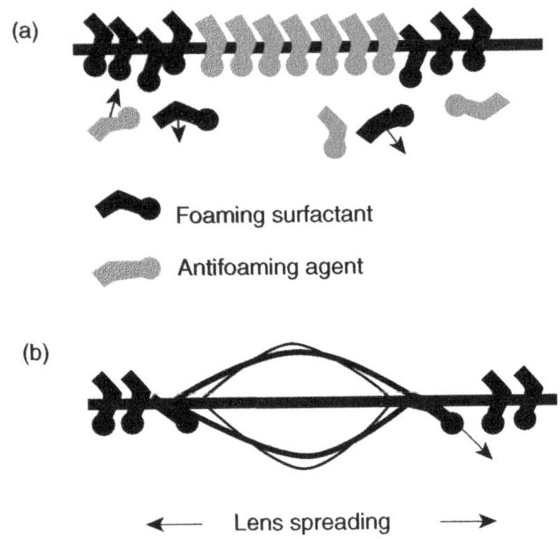

Figure 9.8. Antifoaming agents may act by one or both of the two defoaming mechanisms: **(a)** The antifoam agent molecules may displace the stabilizing surfactant molecules that results in the break down of bubbles or **(b)** the antifoam agent molecules may displace the stabilizing surfactant molecules by spreading as a lens at the gas–water interface [35].

emulsions; therefore, they can be mixed readily with the aqueous foamy liquid. In this manner, the probability for these tiny emulsion droplets of defoamers used at very low concentrations to undergo coalescence via droplet collision is very low. Thus, the function of antifoaming agents can be fully realized.

Typical defoamers include surfactants with very low HLB values, alkyl alcohols ($C_nH_{2n+1}OH$, $n = 6$–10), pine oil, mineral oil, and silicone-based antifoaming agents. Among these materials, silicone-based defoamers are the most effective in suppressing the formation of foams [39]. This is because polydimethylsiloxanes combine two desired physical properties seldom found together, that is, involatility and low surface tension. Furthermore, they are chemically inert and insoluble in both water and lubricating oil. Polydimethylsiloxanes are effective in inhibiting the formation of foams in concentrations in the range of 10 ppm or less, whereas other antifoaming agents are often used in the range of 100–1000 ppm of the bulk fluid. However, as a result of the poor compatibility of the silicone-based antifoaming agent in the coating formulation, surface defects such as craters or fisheyes may occur in the coating films if relatively high levels of silicone-based antifoaming agents are used or there is inadequate incorporation.

9.5 WETTING

Wetting of substrates is defined as the displacement of adsorbed air at the substrate surface by a liquid coating or ink material. Satisfactory wetting pro-

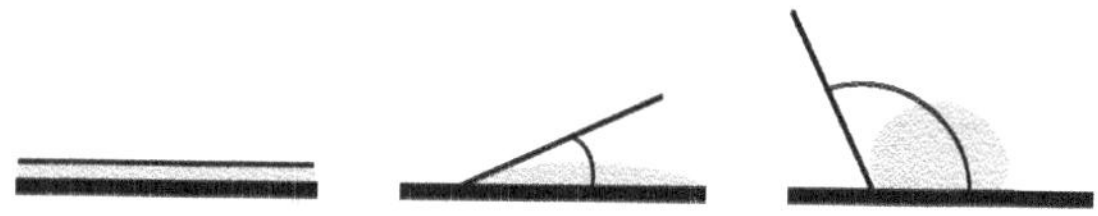

Figure 9.9. A drop of a coating or printing material placed on a solid substrate can ultimately take one of three forms: (a) a thin, uniform polymer film ($\theta = 0°$); (b) a convex lens with a section less than the radius of curvature ($0° < \theta < 90°$); or (c) a convex lens with a section greater than the radius of curvature ($\theta > 90°$).

cesses are essential for successful coating application and printing. This issue is especially crucial for the application of water-based polymer systems onto substrates with low surface energy or substrates contaminated by dirty particles or grease. Even for substrates with relatively high surface energy, wetting may become an important factor when highly dynamic application processes (e.g., fast-running printing and roller-coater application processes) are involved. Under these circumstances, the wetting processes must be very fast in order to obtain satisfactory polymer films. Failing to wet the substrates results in surface defects such as crazing, crawling, or even poor adhesion.

There are two aspects of the water-based coating and printing systems that must be considered in the wetting and spreading processes, that is, equilibrium thermodynamics and kinetics. The former subject will be the focus of this section.

One of the primary characteristics of coating emulsion polymer-based products onto substrates is the contact angle (θ). When a drop of a water-based coating or printing material is placed on a solid substrate, the coating fluid will either spread across the substrate surface to ultimately form a thin, relatively uniform film ($\theta = 0°$, Figure 9.9a) or spread to a limited extent but remain as a discrete drop on the surface (Figures 9.9b and 9.9c). The smaller the contact angle, the better the wetting of the substrate surface by the coating material. As rules of thumb, some general guidelines can be established.

(a) Solid substrates with higher surface energy are easier to wet.
(b) Water-based coating and printing materials with lower surface tension (i.e., surface energy) wet substrates more effectively.
(c) Excellent wetting is achieved when the coating material has a much lower surface energy than the substrate.

The static surface tensions of common solvents range from $14\,\mathrm{mM\,m^{-1}}$ for isopentane up to $72\,\mathrm{mN\,m^{-1}}$ for water. Thus, it is not surprising that current water-based coating and printing systems very often exhibit wetting problems on solid substrates with relatively low surface energy. Table 9.2 lists typical values of critical surface tension of wetting at which $\cos\theta = 1$ (σ_c) for a variety of commonly encountered materials [35]. The critical surface tension of wetting

Table 9.2. Typical Critical Surface Tension of Wetting (σ_c) Data for Commonly Encountered Materials [35]

Solid	σ_c (mN cm^{-1})	Solid	σ_c (mN cm^{-1})
Teflon	18	Copper	60
Polytrifluoroethylene	22	Silver	74
Polyvinylidene fluoride	25	Silica (dehydrated)	78
Polyvinyl fluoride	28	TiO_2 (anatase)	92
Polyethylene	31	Graphite	96
Polystyrene	33	Lead	99
Polyvinyl alcohol	37	Tin	101
Polyvinyl chloride	39	Iron	106
Polyvinylidene chloride	40	Iron oxide	107
Polyethyleneterephthalate	43	Silica (hydrated)	123
Nylon 6,6	46	TiO_2 (rutile)	143

is defined as the surface tension of a liquid that would just spread on the substrate surface to give complete wetting. It is not a characteristic property of the substrate alone, but of the solid–liquid combination. However, this concept is very useful in developing a method for characterizing the wettability of a substrate surface.

Two approaches to effectively improve the wetting of solid substrates can be identified:

(a) Increasing the surface energy of substrates via thorough cleaning (removal of oils or other contaminants) and/or surface treatment (e.g., corona pretreatment, flaming, acidic or caustic wash).

(b) Lowering the surface tension of water-based coating and printing systems by using additives termed substrate-wetting agents.

Just like conventional surfactants, a wetting agent possesses both a hydrophilic part and hydrophobic part. Most hydrophilic parts are ionic groups or noionic polyethylene glycols. The hydrophobic parts are normally selected from hydrocarbon chains. The wetting agents containing fluorinated groups or polysiloxane chains are extremely effective in lowering the surface tension of waterborne coating and printing systems at relatively low concentrations. Furthermore, they can impart special physicochemical properties to the coating formulations. It is also crucial to assure that the wetting agents of choice should not result in undesirable side effects such as the interference with intercoat adhesion, enhanced formation of foams, and reduced water resistance. In highly dynamic application processes (e.g., printing), the static surface tension of waterborne coating and printing systems is not the predominant factor that controls the wetting behavior. In this case, the extremely mobile wetting agent species are capable of orienting themselves rapidly at the newly created interfaces during the coating application process. As a consequence, they can serve as effective substrate-wetting additives under highly dynamic printing or coating operations. As mentioned in Chapter 2, the hydrophile–lipophile balance (HLB) values of common wetting agents range approximately from 7 to 9.

9.6 SURFACE MODIFICATIONS

The surfaces of the coated solid substrates are generally required to display satisfactory physicochemical properties such as excellent appearance (e.g., gloss, matting, texture, soft and silky feel, etc.), slip, blocking resistance, abrasion resistance, scratch and mar resistance, water resistance, and chemical resistance. In general, these desirable surface properties cannot be achieved for water-based coating and printing systems in the absence of surface modifying additives.

Physicochemical properties such as wetting (see Section 9.5), leveling, soft and silky feel, scratch and mar resistance, and water repellency are strongly dependent on the surface tension of water-based coating and printing systems. It is quite difficult to achieve superior surface properties mentioned herein because of the relatively high surface tension of water-based coating and printing systems (recall that the surface tension of water is 72 dyn cm^{-1}). Additives such as silicones, waxes, and some surfactants are designed for resolving these inherent drawbacks associated with water-based coating and printing systems.

In addition to glossy polymer films, sometimes a lower gloss or a certain texture for the coated solid substrate surface is desired. The greatly reduced gloss of a surface coating can be achieved by the transformation of the originally smooth surface into a surface with micro-roughness. In this manner, the incident light can be scattered significantly (Figure 9.10). Micron-sized silica particles (2–4%) are commonly used to impart the matting effect to water-based coating and printing systems. These porous, nonspherical silica particles tend to protrude from the coated substrate surface when the coating film continues to shrink during the drying process. As a result, a matte polymer film on the substrate is achieved. Sedimentation of the inorganic silica particles and/or poor rheological properties of water-based matte coating and printing materials can become unacceptable in plant production.

Silica particles are also very effective in reducing the gloss of solvent-based coating and printing systems. However, these silica matting agents are far less efficient in creating the matting effect for high-solid or solvent-free coating systems (e.g., UV-curable coating systems) due to the lack of sufficient shrinkage of the coating film upon drying. One alternative is the addition of polypropylene wax particles (3–5 μm) that give the matting effect via performing the task of protruding at the water–air interface. The polypropylene particles with a lower density tend to float to the surface of the coating film. In addition,

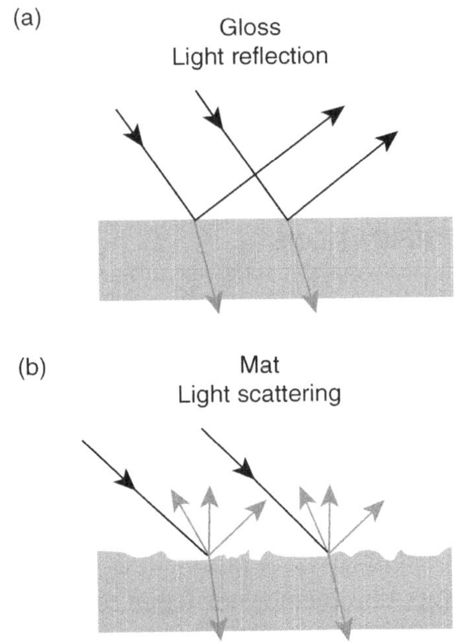

Figure 9.10. An incident light can be either **(a)** reflected by a smooth coating surface or **(b)** scattered to all different directions by a surface with micro-roughness.

significant improvements in the mar and scratch resistance and soft and silky feel can be realized. In a similar manner, polypropylene wax emulsions can be used in water-based coating and printing systems, but the matting efficiency is lower than that of fine silica powders.

Finally, it is interesting to note that combinations of various types of additives generally result in better performance properties of water-based coating and printing systems than one alone. For example, surfactants in combination with waxes significantly enhanced wetting, leveling, blocking resistance, and scratch resistance in water-borne coating and printing systems.

9.7 STABILITY OF LATEX PRODUCTS

When a latex product is frozen, ice crystals tend to undergo phase separation from the colloidal system; therefore, the concentration of polymer particles in the fluid phase continues to increase with the progress of the freezing process. Sooner or later, phase inversion will occur and the probability for the coagulation of polymer particles to take place increases significantly. This is especially true for polymer particles with a glass transition temperature lower than the freezing temperature or with insufficient stabilization by surfactants or protective colloids. Slow freezing of water-based coating and printing systems is much more destructive than fast freezing because larger ice crystals form during the freezing process. Sometimes, strong mechanical agitation may restore the seemingly unstable latex products back to relatively stable state during the freeze–thaw process, though this action is often accompanied by some changes in rheological properties.

Incorporating a small amount of acrylic acid or methacrylic acid (<3%) into the emulsion polymerization systems of acrylate and methacrylate monomers significantly enhances the freeze–thaw stability of the resultant latex products [40]. In addition, the mechanical stability of these latex products is also improved. Latex products stabilized by nonionic surfactants generally exhibit excellent freeze–thaw stability [41]. While the latex particle size, emulsion polymer hardness, surfactants, and protective colloids have some effect on the freeze–thaw stability, adding hydrophilic cosolvents such as ethylene glycol, propylene glycol, and glycerol effectively lowers the freezing point of water and, thus, allows latex products to pass the freeze–thaw stability test. The drawback of using water-soluble cosolvents is the greatly reduced drying speed of the coated polymer films during application and an increase in VOCs. Many nonionic surfactants also help improve the freeze–thaw stability of water-based coating and printing systems.

The freeze–thaw stability of water-based coating and printing systems can be determined by observing how many freeze–thaw cycles they will go through without affecting the basic performance properties such as flocculation of polymer and pigment particles and the degree of increase in viscosity. A freeze–thaw cycle consists of freezing the sample at a temperature of 10°F over a period of 16h, followed by thawing the sample at 75°F for 8h. Most commercially available water-based coating and printing materials need to pass three to five freeze–thaw cycles.

Latex products are sometimes subjected to very high mechanical shear rates during manufacturing, transport, and application. The mechanical stability of water-based coating and printing systems is primarily a function of the stabilizing system used and is dependent on the mechanical efficiency of the envelope surrounding the colloidal particles, the electrostatic surface charge density that enables the particles to repel one another, the density of the particles relative to that of the aqueous medium, and the viscosity.

The probability for emulsion polymer particles to collide with one another is linearly proportional to the velocity gradient experienced in the colloidal system [42]. Vigorous mixing significantly increases the frequency and intensity of collision among the particles and very often impairs the colloidal stability of water-based coating and printing materials. As a consequence, undesired coagulation of colloidal particles may take place. At constant molar concentration of a series of nonionic surfactants [e.g., nonylphenol polyethoxylate with an average of n monomeric units of ethylene oxide per molecule ($C_9H_{19}—C_6H_4—O—(CH_2CH_2—O)_n—H$)], the mechanical stability of the "hairy" polystyrene latex particles increases with increasing the polyethylene oxide chain length (n). This is because the thickness of the adsorbed nonionic surfactant layer surrounding the polymer particle increases with increasing the polyethylene oxide chain length. The thicker the adsorbed nonionic surfactant layer, the larger the steric repulsion force between two colliding particles (i.e., the better the mechanical stability of latex particles), as discussed in Chapter 2, Section 2.3.3. Furthermore, addition of a small amount of anionic surfactants such as fatty acid soaps ($C_nH_{2n+1}COOK$, where $n = 11$ is the optimum), alkyl sulfates ($C_nH_{2n+1}OSO_3Na$, where $n = 10$ is the optimum), and alkyl sulfonates ($C_nH_{2n+1}SO_3Na$, where $n = 10$ is the optimum) into natural rubber latex products significantly improves their mechanical stability.

As rules of thumb, the following statements regarding the mechanical stability generally hold valid.

(a) Latex products stabilized by protective colloids (e.g., hydroxyethyl cellulose, polyvinyl alcohol, starch, alkali-soluble polymers, etc.) exhibit excellent mechanical stability.
(b) A latex product with a smaller particle size shows better colloidal stability toward intensive mechanical agitation than that with a larger particle size. This is simply due to the fact that the frequency of collision among the colloidal particles is proportional to the particle size to the third power [42]. This factor would accelerate the particle coagulation process, because latex particles start to lose their colloidal stability and then grow larger and larger. A relatively small number of larger particles in the original latex will also decrease shear stability for the same reason.
(c) The mechanical stability of latex products decreases with increasing temperature.

Standard high-speed stirrers such as Waring Blendor, Atomix, and Hamilton–Beech may be employed for the mechanical stability test. After stirring the sample for a prescribed period of time, any filterable solids are collected by a 100- or 250-mesh screen, weighed, and then expressed as the percentage of the original sample weight.

Latex products may display colloidal instability when stored in warehouses or transported in containers at relatively high temperature (up to 60 °C or even higher). This is because the viscosity of latex products decreases with increasing temperature. Thus, the tendency for unstable polymer particles with a relatively higher density compared to water to undergo coagulation and then sedimentation is greatly increased. One possible mechanism responsible for this instability problem is the greatly reduced stabilization effects provided by surfactants at relatively high temperature. For example, the degradation of the sulfonate-containing anionic surfactants by hydrolysis may occur according to the following reaction [43]:

$$C_{18}H_{37}C_6H_4SO_3Na + H_2O \rightarrow C_{18}H_{37}C_6H_5 + NaHSO_4 \quad (9.4)$$

This degradational reaction can be detected by the continuous increase of surface tension. As a result of the decreased concentration of effective surfactants, latex particles start to lose their colloidal stability upon heating. With nonionic surfactants, for example, the hydrogen bonding strength between the hydrophilic part [i.e., the polyethylene oxide ($-(CH_2CH_2)_nO-$) chain] and water molecules (i.e., the water solubility of surfactant) decreases with increasing temperature. Thus, the adsorbed polyethylene oxide chains surrounding the latex particle shrink and then become less effective in stabilizing the colloidal dispersion (see Chapter 2, Section 2.3.3). Adequate selection of surfactants in preparing emulsion polymer products may alleviate the heat stability problem.

An appreciable dried polymer film (skin) is very often observed on the top of a latex product during storage at relatively high temperature. This is attributed to the inevitable evaporation of water from latex products. Addition of a small amount of hydrophilic cosolvents such as propylene glycol can increase the boiling point of the aqueous medium and, thus, retard the formation of the undesired dried polymer film.

REFERENCES

1. D. Sardar, S. V. Radcliffe, and E. Baer, *Polym. Eng. Sci.* **8**, 290 (1968).
2. A. M. Sookne and M. Harris, *Ind. Eng. Chem.* **37**, 478 (1945).
3. P. J. Flory, *J. Am. Chem. Soc.* **67**, 2048 (1945).
4. P. J. Flory, *Ind. Eng. Chem.* **38**, 417 (1946).
5. E. J. Lawton, J. S. Balwit, and A. M. Bueche, *Ind. Eng. Chem.* **46**, 1703 (1946).
6. L. E. Nielsen, *Mechanical Properties of Polymers and Composites*, Vol. 2, Marcel Dekker, New York, 1974, Chapter 5.
7. J. A. Prane, *Introduction to Polymers and Resins*, Federation Series on Coatings Technology, Federation of Societies for Coatings Technology, Philadelphia, 1986.
8. L. W. Hill, *Mechanical Properties of Coatings*, Federation Series on Coatings Technology, Federation of Societies for Coatings Technology, Philadelphia, 1987.
9. S. Ikeda, *Prog. Org. Coat.* **1**, 205 (1973).
10. A. Zosel, *Prog. Org. Coat.* **8**, 47 (1980).
11. J. W. Goodwin, *Colloids and Interfaces with Surfactants and Polymers—An Introduction*, John Wiley & Sons, West Sussex, 2004, Chapter 5.
12. A. Ya. Malkin and A. I. Isayev, *Rheology: Concepts, Methods, and Applications*, ChemTec Publishing, Toronto, 2006.
13. C. S. Chou, A. Kowalski, J. M. Rokowski, and E. J. Schaller, *J. Coat. Technol.* **59**, 93 (1987).
14. A. Kowalski, J. J. Wilczynski, R. M. Blankenship, and C. S. Chou, UK Patent Application, GB, 2194543 A, assigned to Rohm and Haas Company, Philadelphia, March 9, 1988.

15. J. E. Schaller, *Surface Coat. Australia* **22**, 6 (1985).
16. J. E. Glass, *J. Coat. Technol.* **50**, 56 (1978).
17. D. H. Blake, *J. Coat. Technol.* **55**, 33 (1983).
18. S. G. Croll and R. L. Kleinlein, in *Water-Soluble Polymers: Beauty with Performance*, J. E. Glass (Ed.), Advances in Chemistry Series 213, American Chemical Society, Washington, D.C., 1986, Chapter 17, p. 333.
19. A. C. Sau, in Polymers in *Aqueous Media: Performance through Association*, J. E. Glass (Ed.), Advances in Chemistry Series 223, American Chemical Society, Washington, D.C., 1989, Chapter 18, p. 343.
20. J. W. Goodwin, R. W. Hughes, C. K. Lam, J. A. Miles, and B. C. H. Warren, in Polymers in *Aqueous Media: Performance through Association*, J. E. Glass (Ed.), Advances in Chemistry Series 223, American Chemical Society, Washington, D.C., 1989, Chapter 19, p. 365.
21. G. D. Shay, in Polymers in *Aqueous Media: Performance through Association*, J. E. Glass (Ed.), Advances in Chemistry Series 223, American Chemical Society, Washington, D.C., 1989, Chapter 25, p. 457.
22. S. Lesota, E. W. Lewandowski, and E. J. Schaller, in Polymers in *Aqueous Media: Performance through Association*, J. E. Glass (Ed.), Advances in Chemistry Series 223, American Chemical Society, Washington, D.C., 1989, Chapter 26, p. 543.
23. K. L. Hoy and R. C. Hoy, United States Patent 4426485, assigned to Union Carbide Corporation, 1984.
24. R. D. Jenkins, *The Fundamental Thickening Mechanism of Associative Polymers in Latex Systems: A Rheological Study*, PhD Dissertation, Department of Chemical Engineering, Lehigh University, Bethlehem, PA, 1990.
25. W. E. Dillon, D. A. Matheson, and E. B. Bradford, *J. Colloid Sci.* **6**, 108 (1951).
26. G. L. Brown, *J. Polym. Sci.* **22**, 423 (1956).
27. J. W. Vanderhoff, H. L. Tarkowski, M. C. Jenkins, and E. B. Bradford, *J. Macromol. Chem.* **1**, 361 (1966).
28. G. Mason, *Br. Polym. J.* **5**, 101 (1973).
29. Z. W. Wicks, Jr., *Film Formation*, Federation Series on Coatings Technology, Federation of Societies for Coatings Technology, Philadelphia, 1986.
30. S. T. Eckersley and A. Rudin, *J. Coat. Technol.* **62**, 89 (1990).
31. S. T. Eckersley and A. Rudin, *J. Appl. Polym. Sci.* **53**, 1139 (1994).
32. D. R. Gehman, J. M. Owens, and R. E. Zdanowski, United States Patent 4150005, assigned to Rohm and Haas Company, 1979.
33. M. J. Devon, J. L. Gardon, G. Roberts, and A. Rudin, *J. Appl. Polym. Sci.* **39**, 2119 (1990).
34. S. Omi, T. Kohmoto, and M. Iso, *Polym. Int.* **30**, 499 (1993).
35. D. Myers, *Surfaces, Interfaces, and Colloids, Principles and Applications*, 2nd ed., Wiley-VCH, New York, 1999.
36. A. W. Adamson, *Physical Chemistry of Surfaces*, 5th ed., John Wiley & Sons, New York, 1990, Chapter XIV.
37. S. Ross, A. F. Hughes, M. L. Kennedy, and A. R. Mardoian, *J. Phys. Chem.* **57**, 684 (1953).

38. J. J. Bikerman, *Foams*, Springer-Verlag, New York, 1973.

39. L. A. Rauner, Antifoaming agents, *Encycl. Polym. Sci. Technol.* **2**, 164 (1964–1972).

40. W. R. Conn, B. B. Kline, and W. O. Prentiss, United States Patent 2795634, assigned to Rohm and Haas Company, 1957.

41. I. I. Eliseeva, T. B. Gonsovskaya, and R. E. Neiman, *Kolloid Zh.* **32**, 856 (1970).

42. P. C. Hiemenez, *Principles of Colloids and Surface Chemistry*, 2nd ed., Marcel Dekker, New York, 1986.

43. H. Warson and C. A. Finch, *Applications of Synthetic Resin Latices, Vol. 1, Fundamental Chemistry Of Latices and Applications in Adhesives*, John Wiley & Sons, New York, 2001, Chapter 4.

INDEX

Printed and bound by CPI Group (UK) Ltd, Croydon, CR0 4YY
16/09/2026
14963459-0001